CIRCULAR STATISTICS IN R

Circular Statistics in R

ARTHUR PEWSEY
University of Extremadura

MARKUS NEUHÄUSER
RheinAhrCampus

GRAEME D. RUXTON
University of St Andrews

Great Clarendon Street, Oxford, OX2 6DP,
United Kingdom

Oxford University Press is a department of the University of Oxford. It furthers the University's objective of excellence in research, scholarship, and education by publishing worldwide. Oxford is a registered trade mark of Oxford University Press in the UK and in certain other countries

First Edition published in 2013

Published in the United States of America by Oxford University Press
198 Madison Avenue, New York, NY 10016, United States of America

British Library Cataloguing in Publication Data
Data available

Library of Congress Control Number: 2013940576

ISBN 978–0–19–967113–7

AP: To Lucía and the memory of my father
MN: To Louis, Emilia, Victoria, and Lennart
GDR: To Katherine and Hazel

PREFACE

As explained in the appendix, just six books providing in-depth treatments of circular statistics have previously been published. There were various motivating factors which prompted us to add this book to that short list. First, the last book published on the topic appeared over ten years ago, and much has changed, in statistics in general and circular statistics in particular, in that time. We felt the time was right to offer a book that provided readers with the background to, and the functionality to apply, traditional as well as more recently proposed methods for analysing circular data. In particular, we stress the use of likelihood-based and computer-intensive approaches to inference and modelling, and distributions that are capable of modelling features such as asymmetry and varying levels of kurtosis that are often exhibited by circular data. Also in recent years, the **R** programming language and environment has become increasingly popular amongst those wishing to implement new statistical techniques quickly and reliably. The appearance of its excellent **circular** package confirmed to us that **R** was the ideal language in which to program functions to implement the different techniques which users would find easy to use.

When writing this book we have sought to serve a number of potential readerships. First and foremost, we wanted to offer a short but authoritative guide to the analysis of circular data for scientists who would not necessarily describe themselves as statisticians, but who have circular data to analyse and want to do so as effectively as possible. Although we make extensive use of **R**, the book is not simply a manual about implementing techniques in **R**. We would hope that it provides useful guidance on the statistical methodologies available to explore circular data, regardless of the computer package that the user adopts to implement those ideas. That said, we picked **R** not only because of its strong and still-growing popularity amongst scientists, but because we feel that it offers a powerful toolkit for effective exploration of circular data. We hope that this book showcases that power and helps the reader to fully exploit it. Lastly, we hope this book will also be of interest to those scientists who would describe themselves as statisticians. There are many interesting methodological challenges still to be resolved in circular statistics; we highlight some of them throughout the text in the hope of inspiring more statisticians to apply their skills in this field.

Arthur has devoted much of the last 15 years of his life to research in the field of circular statistics, Markus is an expert in computer-intensive statistical methods, and Graeme is a biologist with particular interest in making modern statistical techniques accessible to broad user groups. Hopefully, together, we have managed to provide just the book on circular statistics that you are looking for. If not, or if you spot any errors or disagree with anything we have written, we would very much appreciate an e-mail.

Arthur Pewsey, Cáceres, Spain (apewsey@unex.es)
Markus Neuhäuser, Remagen, Germany (neuhaeuser@rheinahrcampus.de)
Graeme D. Ruxton, St Andrews, Scotland (graeme.ruxton@st-andrews.ac.uk)
March 2013

ACKNOWLEDGEMENTS

First, we should thank Clare Charles and Keith Mansfield at Oxford University Press for their support and advice throughout the production of this book. Clare also gathered a number of insightful reviews of our initial proposal, and we thank reviewers for their wise advice towards improving the book. We thank Gandhimathi Ganesan of Integra Software Services Pvt Ltd and copy-editor Mike Nugent for helpful and skilled production of the final layout of the book. As mentioned in the preface, throughout the book we make extensive use of **R**. We would like to acknowledge the selfless work of many that continues to be invested in improving this wonderful statistical and graphical environment. In particular, we would very much like to acknowledge our considerable debt to the authors of its exceptional **circular** package: Ulric Lund and Claudio Agostinelli. Arthur would also like to thank his co-authors Toshihiro Abe, Chris Jones, Shogo Kato, Bill Reed and Kunio Shimizu for sharing their thoughts on circular statistics with him, and Toby Lewis who has been a great inspiration not only for this book but also for no less than three of the other books published on circular statistics. Finally, we would like to thank Marie-Therese Puth for having independently checked the final draft of the book and the **R** code for typographical errors. This work builds on the existing literature on circular statistics, and we have endeavoured throughout to properly cite existing books and papers; but if any author feels that we have not given them full and fair acknowledgement, please let us know so we can make amends.

CONTENTS

1

Introduction

1.1 What is Circular Statistics?

The term *circular statistics* refers to a particular branch of the discipline of statistics that deals with data that can be represented as points on the circumference of the unit circle. Data of this type are themselves referred to as being *circular*, a term used to distinguish them from the usual *linear* data that we are more used to. More formally, we say that the *support* for circular data is the unit circle (as opposed to the real line which is the support for linear data).

Examples of circular data include directions measured using instruments such as a compass, protractor, weather vane, sextant or theodolite. It is usual to record such directions as angles expressed in degrees or radians measured either clockwise or counterclockwise from some origin, referred to as the *zero direction*. The requirements to specify the position of the origin and the direction taken to be positive do not arise for data on the real line; the origin is 0, values to the left of 0 are negative and those to the right are positive. For circular data, each angle defines a point on the circumference of the unit circle, just as each value of a linear variable defines a point on the real line. As the absolute value of a linear variable increases we move further away from the origin. So, on the real line, a value of 360 is relatively close to a value of 355 but relatively far from the origin. The situation is very different for circular variables. Whilst an angle of 355° corresponds to a point on the circumference of the unit circle that is close to that corresponding to 360°, the angles 0° and 360° define the exact same point. It is this periodic nature of circular data that forces us to abandon standard statistical techniques designed for linear data in favour of those which respect the periodicity of circular data.

As an illustration of what can go wrong if we treat circular data as being linear, suppose we measured the directions of flight, clockwise from north, of homing pigeons released at a certain location. If the angles measured for four birds were 10°, 20°, 340° and 350° then commonsense tells us that the birds generally fly approximately northwards. However, the arithmetic mean of these angles is 180°—directly due south! In Chapter 3 we will introduce statistical summaries that take account of the periodic nature of circular data.

Whilst measured directions recorded as angles constitute one type of circular data, not all circular data are necessarily initially measured or recorded as angles. The key to understanding this point is the periodicity of circular data. Consider, for example, the time of day measured on a 24-hour clock. The times 0:00 and 24:00 both correspond to midnight, and 1:00 and 23:00 both define times one hour from midnight. Each time corresponds to a position of the hour hand on a 24-hour clock, and each such position can be converted to an angle measured in degrees by multiplying the time in hours by 360/24. Those angles can then be used to define points around the circumference of the unit circle. Data of this form could represent the times during the day of the onset of menstruation, for example, or those of violent attacks by enemy troops. Other circular variables are related to the time of the year. For instance, we might be interested in the occurrence throughout the year of Japanese earthquakes with a magnitude of 5 or more on the Richter scale. Then, the mighty earthquake that struck on 11 March 2011 was very close, in terms of the time of the year, to one of magnitude 6.9 that occurred on 14 March 2012. Data of this type can be converted to angles measured in degrees by multiplying the time of the year in days since 0:00 hours on 1 January by 360/365 (if we are prepared to ignore leap-years), and subsequently represented as points on the circumference of the unit circle.

From the examples already referred to above, it is clear that circular data will be of interest in many contexts. Further examples include the bonding angles of molecules, the direction of the wind measured at a wind farm at 12:00 each day, the times during the day of cyber-attacks at an intelligence centre, and the incidence throughout the year of measles, lightening strikes on a major city, or solar flares. Other applications from astronomy, geology, medicine, meteorology, oceanography, physics and psychology are referred to in Fisher (1993) and Mardia and Jupp (1999). As Morellato *et al.* (2010) discuss, variables that characterize the phenology of species, such as flowering onset during the year, are of great interest to biologists. Many of these examples illustrate the importance of circular statistics in environmental and climate-change analysis. Circular statistics has also been applied recently by Mardia *et al.* (2007) and Boomsma *et al.* (2008) in the areas of bioinformatics and proteomics.

The reference, or 'zero', point, such as 1 January, midnight or north, is an arbitrary human construct that generally does not relate well to the underlying drivers of the system under study. We might, by convention, label 1 January as day one and 31 December as day 365, but in terms of measles incidence in a large metropolitan area, or the prevailing wind direction at a weather station, we might expect the values recorded on these days to show strong commonality. Near-global agreement on the timing of New Year is a relatively recent construct reflecting the pervading influence of Western culture; there is no astronomical reason why it should be so.

As circular data are ultimately represented as angles or, equivalently, as points on the circumference of the unit circle, it should be no surprise that mathematical results for unit vectors involving trigonometric functions figure within many of the methods presented in the following chapters.

1.2 What is R?

R (**www.R-project.org**) is a software language and environment designed for statistical computing and graphics production. With its extensive functionality and object-orientated philosophy, in recent years it has become *the* platform with which to develop new statistical techniques. Some of the main reasons why **R** has become so incredibly popular are, undoubtedly, that it is free and open source, as well as the fact that users have developed hundreds of packages coded in **R** with which to perform established as well as innovative statistical techniques. Like S-Plus (www.tibco.com), **R** is based on the **S** language developed at Bell Laboratories. Nowadays, **R** is widely used not only by statisticians, but also in fields such as physics, chemistry, sociology and, notably, biology. Although **R** comes without any warranty, many of its packages have been written by experts in their specific fields. Moreover, the core language has a long history and has been widely used and tested, and thoroughly debugged. Being command-driven, **R**'s learning curve is initially relatively steep but, because of its functionality, flexibility and user support, novices soon find it relatively easy to write their own code.

1.3 Getting Started with R

The **R** software and all of its user-contributed packages are available at the **cran.r-project.org** website. In order to download **R** from the nearest site (or *mirror*) you need to search the 'comprehensive **R** archive network' (or **CRAN** for short). The time needed to download and install the core **R** software will depend on the specifications of your computer and Internet connection. To speed things up it is generally best to use a physical Internet connection rather than Wi-Fi.

Once you have installed **R**, to boot the software double click on the icon that will have been pasted on the dashboard of your computer during the installation process. A relatively spartan graphical user interface (GUI) will then open and, within it, the **R** command window. This is when your learning curve kicks in!

R has a detailed and extensive web-based help facility. To obtain help on a function, **t.test** for instance, simply type **help(t.test)**, or **?t.test**, on the command line. In response, a webpage explaining how Student's t-test is performed within **R** will open. At the very end of each such page, various examples of the use of the function concerned are presented. Copying and pasting those examples into the command window is generally a good way to learn how **R** works.

Of course, when you start using **R** you will not know what functionality **R** has. To find out what help is available within **R** for a given theme or topic, for instance 'test', simply type **??test**. In response, **R** opens a webpage with details of, and links to, all those functions available with the word 'test' in their description. Further help is available from the **cran.r-project.org** website in the form of various manuals that provide an introduction to, and overview of, **R** and its functionality. Should you need extra basic information about **R** you may find introductory texts such as Adler (2010), Kabacoff (2011), Crawley (2012) and Ekstrom (2011) helpful. An excellent web-based aid for beginners is Robert Kabacoff's **QuickR** website (**www.statmethods.net**).

1.4 R's Circular Package

Throughout this book we make extensive use of Ulric Lund and Claudio Agostinelli's excellent **circular** package (**https://r-forge.r-project.org/projects/circular/**), written to perform basic manipulation of, and statistical techniques for, circular data. Here we provide details of how to download the **circular** package and its latest documentation, how to access the data sets available within it, and how its **circular** function works.

The command **library()** produces a descriptive list of all the packages you presently have installed on your computer. Since the **circular** package is not a standard package it is not distributed with the base **R** software. It must therefore be downloaded from one of **CRAN**'s mirrors. When connected to the Internet, packages can be installed on your computer using the command **install.packages()**. Choosing **circular** from the long list of packages that will appear when using this command, you can install the **circular** package onto your computer. To make the **circular** package available within your present **R** session you have to load it using the command **library(circular)**.

Once loaded, you can use all the functions implemented in the **circular** package. You also have access to the data sets that come with the package. One such data set, which we will use later in Chapter 5, is contained in the circular data object **fisherB1c**. You can visualize its data values by simply typing the name of the data set, i.e. **fisherB1c**, on the command line. A description of the data is provided by the help page opened by typing **?fisherB1c**. A descriptive list of all the data sets available in your current session, including all those available within the **circular** package, can be produced using the command **data()**.

The latest documentation for the **circular** package is available from **cran.r-project.org/web/packages/circular/index.html**. Besides the reference manual (**circular.pdf**), the package source code and binaries can also be downloaded from this webpage. One of the basic functions referred to in the **circular** package's reference manual is **plot.circular** (or **plot** for short). To see what this function produces for a circular data object, on the command line type:

```
plot(fisherB1c, pch=16, col="blue", stack=T, shrink=1.2, bins=720, ticks=T)
```

To get a better feel for how this function works, try changing some of its modifiers. For instance: **pch=16** to **pch=1**; **col="blue"** to **col="red"**; **shrink=1.2** to **shrink=1.4**; **ticks=T** to **ticks=F**. The reference manual provides details of other modifiers that can be used to change the plot's appearance. We will say more about the use of the **plot.circular** function in Section 2.2.

The functions available within the **circular** *package* generally assume that any data objects have been prepared using, rather confusingly, its **circular** *function*. Since we will make extensive use of the **circular** function, here we explain the key properties of a data object prepared using it. One such property is its units, specified using the modifier **units** and one of the options "**radians**", "**degrees**" or "**hours**". Radian measure is the default option. Further modifiers are:

- **modulo** which specifies how data values should be remaindered. This modifier has the options "**asis**" (for no change), "**2pi**" and "**pi**". No change is the default. If we set **modulo="2pi"** then values greater than 2π are replaced by the remainder after dividing them by 2π, thus ensuring that all values fall in $[0, 2\pi)$.
- **rotation** which determines the direction of rotation from the origin. It has the options "**counter**" and "**clock**". Counterclockwise is the default.
- **zero** which specifies where the zero direction is located assuming radian measure and counterclockwise rotation from the standard mathematical origin. So the default, **zero=0**, locates the zero direction at the positive horizontal axis. The other commonly used setting (especially for data relating to the time of day) is **zero=pi/2** which locates the zero direction at the positive vertical axis.
- **template** which can be used to specify the values of **modulo, zero** and **rotation** simultaneously. The default is "**none**" and the alternative options are "**clock12**", "**clock24**" and "**geographics**". The first two are used with data measured as times on a 12, and on a 24, hour clock, respectively. The last sets **rotation** to "**clock**" and **zero** to **pi/2**, corresponding to standard geographical angular measurements made from north.

1.5 Web-based R Code and the CircStatsInR Workspace

On the book's website (**http://circstatinr.st-andrews.ac.uk/**) we provide **.txt** files containing the **R** code used in each one of the subsequent chapters. That code can be copied and pasted into **R**'s command window to repeat the analyses we present. Another possibility is to copy those files to others which you can then edit in order to run similar analyses of your own. In the interests of efficiency, it is always a good idea to save any useful code that you develop before terminating an **R** session. In later sessions you can then easily edit the saved code, or simply copy and paste it into the command window to repeat an analysis.

The **CircStatsInR** workspace is also available from the website. It contains all of the new functions that we introduce within the book as well as an extra data object referred to in Chapter 6. We recommend that you download a copy of it to an appropriate directory on your computer. Double clicking on its icon automatically opens its content in **R**. Once you have it open in **R** you can obtain a list of the functions and objects it contains by typing **ls()** on the command line. By using the **CircStatsInR** workspace you will avoid the need to copy and paste the code for the new functions that we introduce within this book. Instead, you can make use of any one of them by simply typing its name followed by the values of its arguments between brackets. As an example, and assuming you have the **CircStatsInR** workspace open and the **circular** package loaded, on the command line type:

```
vMPPQQ(circular(wind), circular(0.3), 1.8)
```

In Section 6.2.3 we will explain what diagrams like the one produced using the above command represent and how they can be interpreted.

1.6 Circular Statistics in Other Software Environments

Oriana (www.kovcomp.co.uk/oriana) is, as far as we are aware (see also Morellato *et al.* (2010)), the only platform dedicated specifically to circular statistics. Other potentially useful software resources include Nick Cox's CIRCSTAT Stata (www.stata.com) modules (http://EconPapers.repec.org/RePEc:boc:bocode:s362501), the CircStat toolbox for use with Matlab (www.mathworks.com) developed by Berens (2009), Nick Fisher's S-Plus software for plotting circular data (www.valuemetrics.com.au/resources005.html) and the S-Plus subroutines for the analysis of circular data provided by Jammalamadaka and SenGupta (2001). The latter were ported to **R** in the **circular** package's predecessor **CircStats**.

All of the methods presented in this book could be programmed in any other statistical software environment. However, because of the reasons referred to in Section 1.2 and the existence of **R**'s **circular** package, we consider that **R** is, by far and away, 'the way to go'. Nevertheless, we feel sure that the book will also be of benefit to those who are happier using other software environments. Hopefully, they will be able to translate any **R** code that we present into software of their own.

1.7 Related Types of Data

This book focuses mainly on data for which the natural support is the unit circle. So-called *axial* data, for which the angles θ and $\theta + \pi$ (radians) are indistinguishable, can be represented as points on the semi-circumference of a semicircle of unit radius. Examples of axial data include the orientations of the major axis of termite mounds, or the angles of slope of different sedimentary layers of an exposed rockface. We briefly consider such data in Section 3.7. Turning to bivariate data, the natural support for joint observations on two circular random variables is the unit torus; and a cylinder with unit radius for joint observations on one circular random variable and one linear one. Models for data on such supports are mentioned briefly in Sections 4.3.17 and 4.3.18, and correlation and regression techniques for use with them are considered in Chapter 8. Further extensions of the unit circle are to the unit sphere and the unit hypersphere. Data on such supports are common in many disciplines. We have not considered statistical techniques for spherical data primarily because, at the time of preparing this book, they were not well-supported in **R**. Those interested in the subject should consult Watson (1983), Fisher *et al.* (1993) and Mardia and Jupp (1999). Circular, axial, toroidal, cylindrical and spherical statistics are all subfields of the over-arching field referred to as *directional statistics*. The final two chapters of Mardia and Jupp (1999) consider data on different manifolds and the related field of *shape analysis*. Dryden and Mardia (1998) provide a more extensive treatment of shape analysis.

1.8 Aims of the Book

When preparing this book we have had various ambitious aims in mind. Firstly, we have attempted to keep its length relatively short so that newcomers to the field can quickly identify the methods available for those statistical issues of most practical relevance. For any given topic we have not sought to provide a comprehensive coverage of all available methods but instead to give guidance on how to perform the most suitable exploratory analysis with the methodology currently available. By so doing, we hope the book will help readers make the best use of their data as well as informed inferences and interpretations of them.

Another major aim has been to promote a modern, computer-based approach to the analysis of circular data, founded upon the use of **R**, its **circular** package and new functions that we introduce in the text. Some strong reasons for choosing **R** were discussed in Sections 1.2 and 1.4. The existence of its **circular** package simplifies the manipulation and basic analysis of circular data considerably. By making the **R** code employed throughout the book available we hope that users will better understand techniques and be in a position to write code and functions to implement extensions of those techniques as well as new methods of their own.

We have also sought to promote statistical models and exploratory methods that have received little or no attention in previously published books on circular statistics. For instance, in Chapter 2 we discuss the use of kernel density estimation and composite plots when representing circular data graphically. Chapter 4 provides details of numerous models for circular data that have been proposed only very recently in the literature. A primary motivation for doing so has been to provide analysts with models capable of describing features such as asymmetry and varying degrees of peakedness that circular data often exhibit. As a consequence, we have devoted relatively little space to the classic model of circular statistics—the von Mises distribution. Another consequence has been that in Chapter 6 we have devoted considerable space to likelihood-based approaches to inference and the related issues of model comparison and reduction. Historically, these themes have commanded little or no space within texts on the subject. Many of the large-sample methods considered in Chapter 5 are based on a result published by Pewsey (2004*a*). Throughout the book we champion the use of computer-intensive resampling methods such as randomization tests and the bootstrap. One particularly novel use of such techniques is in the context of goodness-of-fit testing.

The book is not aimed specifically at statisticians, but has been written to be accessible to those working in a wide range of scientific disciplines. We have attempted to keep any assumptions about the reader's mathematical, statistical or computational know-how to a minimum. Nevertheless, some basic knowledge concerning vectors, trigonometric functions, distribution theory, statistical inference and structured computer programming will make assimilation of the full content of the book easier.

1.9 The Book's Structure and Use

As in any exploratory statistical analysis, it is sensible to represent the data at our disposal graphically before applying any formal inferential techniques. By doing so we will get a better feel for what the data are trying to tell us and will be able to identify any atypical observations. Graphical summaries for use with circular data are considered in Chapter 2. Next it is usual to reduce the complexity of the information contained in the original data to numerical summaries which describe their main features. In Chapter 3 we introduce the measures most commonly used to summarize the main characteristics of circular data.

After summarizing our data graphically and numerically, we will generally be interested in modelling them. In Chapter 4 we provide an introduction to the distribution theory underpinning circular statistics and numerous distributions which provide potential models for circular data. The von Mises distribution, which in many ways is the circular analogue of the normal distribution, is one of those models. However, we also present details of other more flexible models, including the Jones–Pewsey and inverse Batschelet families of distributions which include the von Mises distribution as a special case.

Chapter 5 considers certain basic forms of inference which will generally be of interest during the initial exploratory phase of the modelling process. We begin with tests for the fundamental dividing hypotheses of uniformity and reflective symmetry. We also provide the details of distribution-free methods of inference for certain key population summaries.

Model fitting for a single sample is the focus of Chapter 6. There we consider maximum likelihood based point estimation and confidence interval construction for the von Mises distribution and its Jones–Pewsey and inverse Batschelet extensions. We also describe methods employed in the model comparison, model reduction and goodness-of-fit stages of model fitting.

In Chapter 7 we consider hypothesis tests for situations involving two or more circular samples. Chapter 8 deals with correlation and regression methods for use with toroidal and cylindrical data. The book's single appendix provides details of further reading.

In comparison to many other texts on circular statistics, we have reproduced relatively few tables of critical values for test statistics. This is primarily because the percentage points of many standard sampling distributions are readily available in **R**. Also, the use of the computer-intensive forms of inference referred to above obviates the need for such tables. Neither have we included many data sets. The reason for this is that the analyses we present generally make use of circular data sets that come with **R**'s **circular** package.

After introducing each statistical theme, we present the details of methods that can be used to investigate it. As mentioned above, it has not been our intention to provide a compendium of all those statistical techniques that are available for any one theme. Rather the approach taken is to encourage the reader to identify the most appropriate analysis on the basis of:

- the statistical issue under investigation;
- an exploratory investigation of the data;
- background knowledge of the working of the system from which the data were collected.

We guide the reader through the underlying assumptions of the techniques, their related literature and the details of how they perform in an attempt to clarify the most appropriate form of analysis for any given situation. Such an approach is essential, as sometimes there can be a potentially bewildering array of alternative techniques available.

When presenting examples to illustrate the use of the different techniques we always include the code necessary to implement them in **R**. We also emphasize the correct interpretation of the results obtained during any analyses. Numerical results returned by **R** are generally quoted to between two and four decimal places, depending on the circumstances.

When compiling the index we have often collected terms together in a way which at first you may find counter-intuitive. So, if you want to look for, say, the Watson–Williams test for a common mean direction, you will not find it listed under an entry such as 'Watson–Williams test'. Instead, look first under the major grouping entry 'test', then under its sub-entry 'common mean direction', and finally identify the sub-entry 'Watson–Williams'. In following this route you will be led to the other two tests for a common mean direction considered in the book. Our motivation for using such a structuring was that it naturally leads to the identification of related concepts and topics. Moreover, it does not require the reader to necessarily remember the names of the specific concepts and topics themselves. The grouping entry 'test' is, in fact, the last major level grouping entry. The others are, in alphabetical order: 'CircStatsInR workspace'; 'circular'; 'circular package'; 'data'; 'data sets'; 'distribution'; 'inference'; 'plot'; 'population'; 'R'; 'regression'; 'sample'. We hope you will find the use of these major grouping entries helpful. To get a better feel as to how entries have been collected together under them, we recommend you take time out to briefly scan the index. For other concepts and themes, such as 'bandwidth selection' and 'symmetry' for example, simply look directly for their individual entries ordered alphabetically in the index.

1.10 A Note on Resampling Methods

As mentioned in Section 1.8, throughout the book we make extensive use of computer-intensive resampling methods (see, for example, Manly (2007) and Neuhäuser (2012)) to estimate confidence intervals and the p-values of significance tests. Since those methods use computer-generated pseudo-random number sequences, the results obtained using them will vary between different runs of the same code. To ensure that any differences should be slight, we recommend, following the advice of Manly (2007, page 121), that the methods be applied to the original data and $N_R = 9999$ samples generated using the chosen resampling technique.

When estimating the p-value of a test we use the proportion of the $(N_R + 1)$ values of the test statistic that are at least as extreme as the test statistic value for the original data. We will denote that proportion by $\hat{p}$. As the test statistic value for the original data is included in the calculation of the estimate, the lowest possible value of $\hat{p}$ is $1/(N_R + 1)$. Rather than include the original sample, some authors estimate p-values using resampled samples alone. Our adopted approach will produce very slightly more conservative estimated p-values than those obtained using this alternative approach. Being the proportion of Bernoulli trial 'successes', and appealing to the central limit theorem, a $(1-\alpha)100\%$ confidence interval for the true p-value of a test is given by

$$\hat{p} \pm z_{(1-\alpha/2)}\sqrt{\frac{\hat{p}(1-\hat{p})}{(N_R+1)}}, \tag{1.1}$$

where $z_{(1-\alpha/2)}$ denotes the $(1-\alpha/2)$ quantile of the standard normal distribution. Using basic calculus, it is easy to show that the maximum value of $\hat{p}(1-\hat{p})$ is 0.25 and occurs when $\hat{p}=0.5$. Its minimum value is 0, obtained when $\hat{p}=1$ (as noted above, $\hat{p}=0$ is impossible). Thus the width of (1.1) is greatest when $\hat{p}=0.5$, and 0 when $\hat{p}=1$. When $\hat{p}=0.05$ and $N_R=9999$, for instance, the 95% confidence interval for the true p-value given by (1.1) is, to four decimal places, (0.0457, 0.0543). If necessary, the width of any such confidence interval can be reduced by increasing N_R. When the sample size is very small, for some resampling techniques it will be possible to evaluate the complete sampling distribution of a test statistic. Then $\hat{p}$ will coincide exactly with the true p-value.

2

Graphical Representation of Circular Data

2.1 Introduction

Having entered your data into **R** and checked they are correct, the production of some form of graphical representation of them will generally prove insightful. In this chapter we consider those graphical summaries for displaying circular data supported in **R**'s **circular** packages. We start with the most natural graphical summary for circular data; the raw circular data plot. In such a plot, each data point is represented by a dot, or some other symbol, located around the circumference of the unit circle. As we shall see, there are numerous ways in which such plots can be enhanced. Raw circular data plots display all the fine detail of a data distribution. For exploring its larger scale structure, rose diagrams, kernel density estimates and linear histograms are available. Also, insightful composite plots made up of various circular representations of the data can easily be created in **R**.

2.2 Raw Circular Data Plots

The **R** command **plot** is a generic function. Precisely what it produces depends on the libraries that you have previously loaded and the type of data object to which it is applied. For example, the data object might be a standard one, a time series one (produced using **R**'s **ts** function) or a circular one (produced using the **circular** package's **circular** function referred to in Section 1.4).

In order to appreciate the importance of the data object type, consider the data object **wind** available in the **circular** library. This object contains 310 wind directions, measured clockwise from north in radians, recorded at a meteorological station in the Italian Alps every 15 minutes from 3.00am to 4.00am between 29 January, 2001 and 31 March, 2001. These data were introduced to the circular statistics literature by Agostinelli (2007). Although the data contained in **wind** are angles, **wind** is *not* however a circular data object. It is a standard data object containing 310 numerical values with no extra information to communicate to **R** how the data should be interpreted. Without such information, the functionality of **R**'s **circular** library will not be applied to them. We can produce a circular data object containing the data and additional information reflecting the way the data were measured, as well as plots of the original data object and its circular counterpart, using the following commands:

```
library(circular)
windc <- circular(wind, type="angles",units="radians",template="geographics")
plot(wind, pch=16, xlab="Observation number", ylab="Wind direction (in radians)")

plot(windc, cex=1.5, bin=720, stack=TRUE, sep=0.035, shrink=1.3)
axis.circular(at=circular(seq(0,7*pi/4,pi/4)), labels = c("N","NE","E","SE", "S","SW","W","NW"),
zero=pi/2, rotation='clock', cex=1.1)
ticks.circular(circular(seq(0,2*pi,pi/8)), zero=pi/2, rotation='clock', tcl=0.075)
```

Using the **circular** function, **windc** is defined as a circular data object containing angles measured in radians in a clockwise direction from north. The first use of the **plot** function, with the standard data object **wind**, produces the plot in the left-hand panel of Fig. 2.1 in which the values of the angles corresponding to the wind directions are simply plotted sequentially. For reasons discussed in Section 1.1, this linear plot is not a useful summary as it does not reflect the periodicity of the data, for which, for instance, both 0 and 2π radians correspond to north. The second use of the **plot** function, applied to the circular data object **windc**, produces the raw circular data plot in the right-hand panel of Fig. 2.1. The modifiers **cex=1.5, bin=720, stack=TRUE, sep=0.035** and **shrink=1.3**: change the size of the lettering and solid circle symbols used to represent each data point in the plot; locate each point to the nearest half-degree; stack the solid circles corresponding to multiple observations at the same angle (otherwise the number of data points associated with a given angle would be

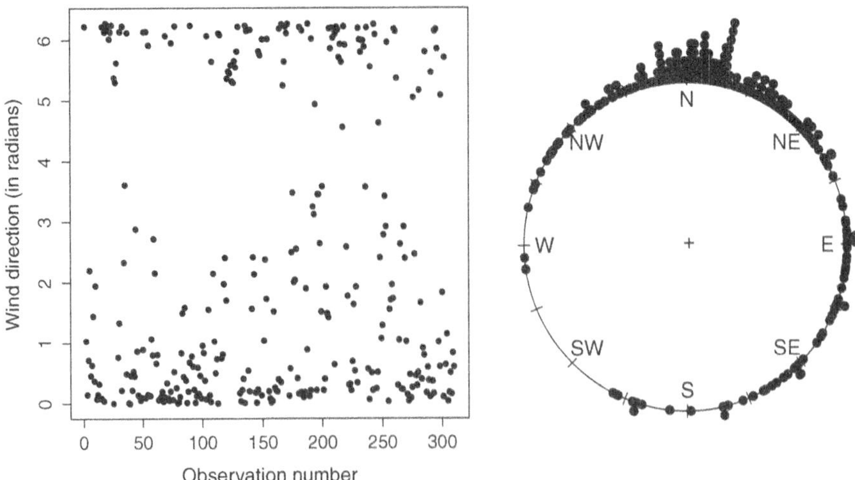

Figure 2.1 Linear plot (left) and raw circular data plot (right) of 310 wind directions, measured in radians clockwise from north, recorded at a meteorological station in the Italian Alps

unclear); increase the separation between the stacked solid circles and reduce the radius of the circle used to represent the unit circle so as to ensure that all of the data points are visible within the plotting window. The function **axis.circular** is then used to add the points of the compass to the plot, starting from the zero direction at north (or $\pi/2$ radians from the mathematical origin corresponding to the positive horizontal axis) and moving in a clockwise direction. Finally, **ticks.circular** adds extra tick marks of specified length to the plot. If the effect of any of these instructions is unclear just try changing the values for the modifiers and run the amended code line by line in order to get a better idea of how they work.

From the right-hand panel of Fig. 2.1 it is clear that the data are concentrated about north and their distribution is asymmetric, there being very few observations in the sector running clockwise from south to west.

The **points** function can be used to display more than one data set within the same raw circular data plot. For example, within the **circular** package the data object **fisherB10c** contains three data sets on the walking directions, measured clockwise from north in degrees, of long-legged desert ants under three different experimental conditions. The following commands plot the three data sets using different plotting symbols and positions relative to the circumference of the unit circle.

```
plot(fisherB10c$set1, units="degrees", zero=pi/2, rotation="clock", pch=16, cex=1.5)
ticks.circular(circular(seq(0,(11/6)*pi,pi/6)), zero=pi/2, rotation='clock', tcl=0.075)
points(fisherB10c$set2, zero=pi/2, rotation="clock", pch=16, col="darkgrey", next.points=-0.1,
cex=1.5)
points(fisherB10c$set3, zero=pi/2, rotation="clock", pch=1, next.points=0.1, cex=1.5)
```

The plotting character drawn is a solid circle when **pch=16**, and an open circle when **pch=1**. The positions of the points relative to the circumference of the unit circle are specified using the **next.points** modifier, and the colour of the points in the interior of the circle using **col="darkgrey"**. The symbols in the resulting plot, portrayed in Fig. 2.2, could have been

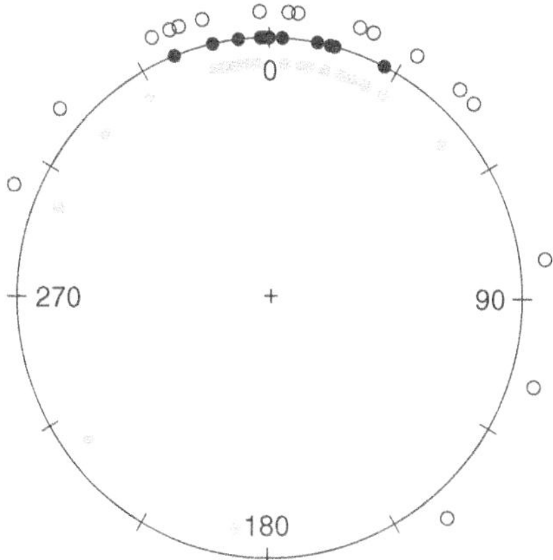

Figure 2.2 Raw circular data plot of the walking directions of long-legged desert ants under three different experimental conditions indicated using different plotting symbols

given more lively colours using, for instance, **col="red"**. Graphical summaries for presentations are particularly enhanced by the use of vivid colours. We will return to the analysis of these data in Chapter 7.

Raw circular data plots are essential when exploring the fine detail of a data distribution. In the next three sections we focus on alternative graphical summaries which can be used to represent its larger scale features.

2.3 Rose Diagrams

Unfortunately, there is no function in **R**'s **circular** library to produce a proper circular histogram (Mardia and Jupp, 1999, Section 1.2.2). A rather basic representation of a circular barchart can be obtained by reducing the value of the **bin** modifier of the **plot** command discussed earlier. However, we do not feel that this approach produces a visually attractive graph. Instead we recommend a close relative of the circular histogram, the rose diagram, in which frequencies are represented by areas of sectors instead of bars. This diagram can easily be drawn using the **rose.diag** function. As an example of its use, the following code, combined with that used to produce the raw circular data plot in Fig. 2.1, adds a rose diagram with 16 class intervals, or bins, in the centre of the unit circle:

```
rose.diag(windc, bins=16, col="darkgrey", cex=1.5, prop=1.3, add=TRUE)
```

The modifier **prop** simply scales the size of all of the segments relative to the outer circle, with larger values increasing the size of the segment, and the default being one. Again it is possible to embellish the basic rose diagram plot in similar ways to those demonstrated with the **plot** function.

The crucial issue to note with a rose diagram is that (unlike linear histograms with rectangular bars) there is not a simple linear relationship between the radius of a segment and its area. As a consequence, it is important to be aware that there are two different conventions in the literature for use with rose diagrams. Here we have adopted the default convention of the **rose.diag** function where the radius of a segment is taken to be the square root of the relative frequency. This is the circular equivalent of the convention used to construct linear histograms. When comparing segments in a rose diagram with this convention, the ratio of the areas of two segments is equivalent to the ratio of the relative frequencies. The other convention is to use a radius which is linearly related to the relative frequency; in this case the radii of the segments, rather than their areas, are what you should compare when assessing the relative frequencies in different segments. This second convention can be implemented by using the modifier **radii.scale = linear**. Whichever one of these conventions you adopt, it is vital that you inform the reader in the figure caption which one you are using and adopt the same convention across any set of figures that might be compared by a reader.

The sectors of the rose diagram in Fig. 2.3 indicate those wind directions that are more or less frequent, as well as highlighting the asymmetry of the data distribution.

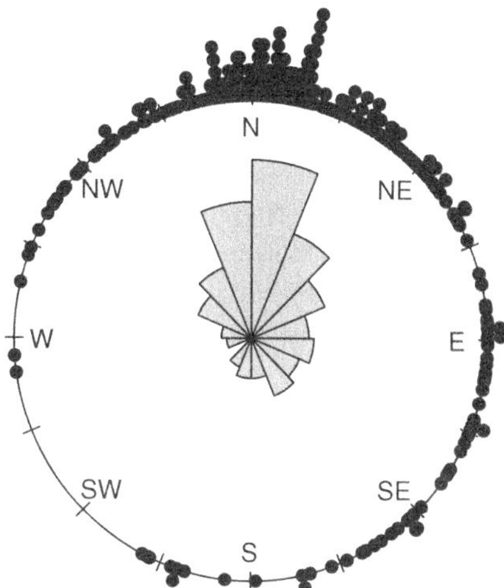

Figure 2.3 Raw circular data plot and, in its centre, a rose diagram for the 310 wind directions. The areas of the sectors in the rose diagram represent the relative frequencies in the 16 class intervals

As with histograms, you have the choice of how many segments of the rose diagram to split your data into. This choice can strongly influence how the graph is interpreted, so we recommend trying a range of values for the number of segments and compare the resulting plots. As a rule of thumb, the square root of the sample size is often used as a reasonable first guess at an appropriate number of segments. For obvious reasons, the values 4, 8, 12, 16, 18, 32 and 36 are popular choices for use with circular data.

2.4 Kernel Density Estimates

In recent years, kernel density estimates have become increasingly popular as graphical representations of circular data. In kernel density estimation the idea is to obtain a nonparametric estimate of the underlying population density.

First a so-called kernel is chosen, which is often a density itself. A kernel can be thought of as a device for spreading the influence of a data point in the close vicinity of that point. For circular data a natural choice of kernel is a von Mises density, the details of which will be given in Section 4.3.8. For now, the von Mises distribution can be thought of as a circular analogue of the normal distribution, and so has the attractive property as a kernel that its density (that is, the influence spreading function) is symmetric and decreases with increasing distance from its centre. A von Mises kernel is the default option for the **circular** package's **density.circular** function. However, other kernels are available.

To calculate the value of a kernel density estimate at a chosen point on the circumference of the unit circle, first identical copies of the chosen kernel are centred at each one of the n observed data points. The values taken by the n kernels at the chosen point are then summed and divided by n.

As with the normal density, the spread of the von Mises density is controlled by a single parameter. However, unlike the standard deviation of the normal distribution, σ, the spread of the von Mises distribution is controlled by a so-called *concentration* parameter, $\kappa > 0$. The larger the value of κ the greater the concentration, and hence the lower the spread, of the distribution. More generally, the choice of the spread parameter that produces the best estimate of the underlying density is referred to as the bandwidth selection problem. When the kernel used with the **density.circular** function is von Mises, the bandwidth value, specified through **bw**, is the value of the concentration parameter κ. Thus, large values of the bandwidth produce spiky kernel density estimates with more modes than the very smooth ones produced with values of κ close to 0.

In an attempt to informally address the bandwidth selection problem, common practice is to produce multiple kernel density estimates with different bandwidths and choose the one which seems to describe the main features of the data best, being neither too spiky nor overly smooth. To illustrate this approach, we can add different kernel density estimates, represented by different line types, to the plot in Fig. 2.3 using the following additional commands:

```
lines(density.circular(windc, bw=75), lwd=2, lty=2)
lines(density.circular(windc, bw=40), lwd=2, lty=1)
lines(density.circular(windc, bw=10), lwd=2, lty=3)
```

Note that the **lwd** and **lty** modifiers of the **lines** command control the width (the default value being **lwd=1**) and type of the lines plotted. As is evident from the left-hand panel of Fig. 2.4, a bandwidth of 10 produces what seems to be the overly smooth density estimate represented by the dotted line type (**lty=3**), with important information about the underlying density being lost. To the other extreme, the density estimate for a bandwidth of 75, delimited by the dashed line type (**lty=2**), appears to be somewhat undersmoothed. So this choice of bandwidth is arguably too high. The solid density estimate (**lty=1**), corresponding to a bandwidth of 40, seems to summarize the main features of the population density best, being neither too rough nor overly smooth. This is the single density estimate included in the less cluttered diagram on the right-hand panel of Fig. 2.4. This final composite plot includes the fine detail of the data themselves, a summary of their gross features, through the rose diagram, and what appears to be a reasonable estimate of the underlying population density. We recommend the use of such composite plots as multilayered summaries of circular data communicating different levels of information.

Rather than simple trial and error, **R**'s **circular** library's **bw.cv.mse.circular**, **bw.cv.ml.circular** and **bw.nrd.circular** functions provide more objective ways of selecting the bandwidth. The first two are based on results from Hall *et al.* (1987) and use cross-validation and minimization with respect to mean squared error loss and Kullback–Leibler loss, respectively, to select the bandwidth. The third uses a rule of thumb proposed by Taylor (2008) that assumes the underlying population to be von Mises and uses a von Mises kernel. Applying these different methods to the **wind** data leads to bandwidth selections of 50, 50 and 11, respectively. Since von Mises distributions are symmetrical and this data set is clearly not, the third bandwidth can be considered unreliable. The other values are relatively close to the value of 40 that we suggested earlier.

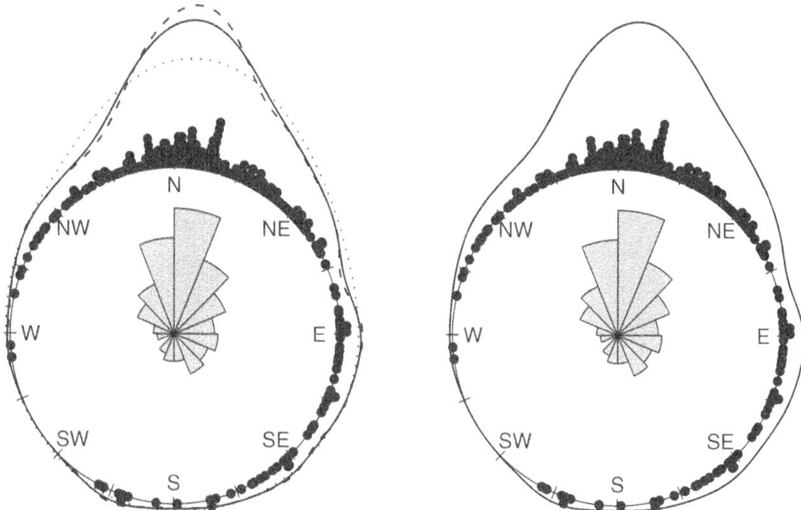

Figure 2.4 Raw circular data plots and rose diagrams of the 310 wind directions together with (left) three kernel density estimates with bandwidths of 10 (dotted), 40 (solid) and 75 (dashed), respectively, and (right) a single kernel density estimate of bandwidth 40. The areas of the sectors in the rose diagram represent the relative frequencies in the 16 class intervals

Recently, Oliveira *et al.* (2012) proposed a new plug-in bandwidth selector based on the use of finite mixtures of von Mises densities. Their simulation results suggest their new approach is highly competitive with other existing bandwidth selection procedures. It, together with various other nonparametric methods for circular data, are implemented in the soon to be released **NPCirc** package.

2.5 Linear Histograms

As proper circular histograms are not available in the **circular** library, what would initially appear to be an appealing alternative would be to produce a standard (linear) histogram of the data. When producing a linear histogram of circular data we effectively cut the circumference of the circle at a specified point and associate the two ends of the cut circle with the two end points of the linear histogram. Consequently, their interpretation is potentially hindered because the periodicity of the data is lost. Implicitly it is assumed that the reader will consciously equate both extremes of the histogram as corresponding to the same point. In effect the reader is required to use some mental gymnastics to wrap the linear histogram back onto the circumference of the unit circle. This is clearly expecting a lot of novices to the field of circular statistics. Further, the issue as to where to cut the unit circle is an especially delicate one.

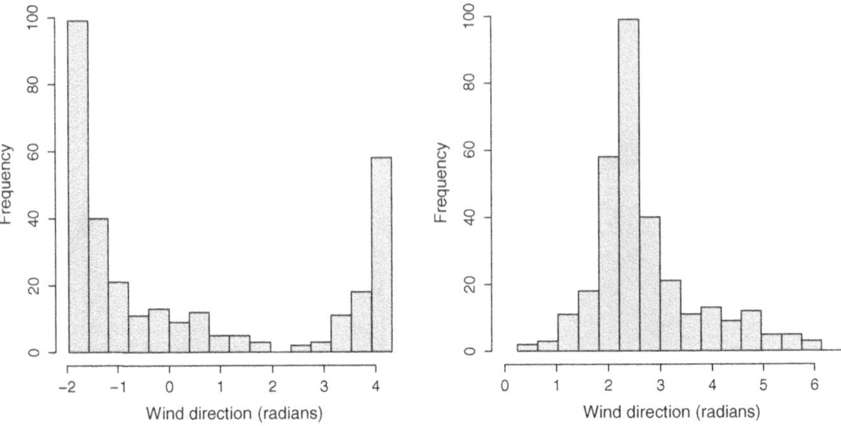

Figure 2.5 Linear histograms of the 310 wind directions with (left) the cut-point of the circle at 0 (north) and a range of $[0, 2\pi)$, and (right) the cut-point of the circle at $2\pi - 5\pi/8$ and a range of $[-5\pi/8, 2\pi - 5\pi/8)$

In order to illustrate these issues, consider the linear histogram in the left-hand panel of Fig. 2.5 of the data in the (standard) data object **wind** (not the circular data object **windc** because the latter cannot be used as the argument of **R**'s **hist** function), produced using the command:

```
hist(wind, main=" ", xlab="Wind direction (radians)", ylab="Frequency", breaks= seq(from=0, to=
2*pi, by=pi/8), col="grey", xlim=c(0,2*pi))
```

What most of the modifiers used within the **hist** function do should, at this stage, be fairly obvious. The **breaks** modifier controls the class intervals of the histogram, there being 16 of them, of equal length, specified using the **seq** function. Without taking into consideration the two important issues discussed above, an initial erroneous interpretation of the histogram might be that the data distribution is bimodal. However, this impression is a consequence of the circle having been cut at an inappropriate place; namely at 0, corresponding to north, around which most of the data are closely distributed. In order to interpret the data distribution correctly, the reader is required to equate the values 0 and 2π. In terms of the linear histogram, this effectively requires the reader to mentally shift that part of the histogram between $2\pi - 5\pi/8$ and 2π to the left of the bar starting at 0. For many, this will be too much to expect.

Clearly, a better place to have cut the circle would have been $2\pi - 5\pi/8$ where least data are concentrated. In order to produce a histogram ending at this point, we first need to edit the data slightly. The necessary changes can be made using the following **R** code:

```
n <- length(wind) ; cutpoint <- 2*pi-(5*pi/8) ; windshift <- 0
for (j in 1:n) { if (wind[j] >= cutpoint) { windshift[j] <- wind[j]-2*pi }
else { windshift[j] <- wind[j] } }
```

In the first line of this code, the **length** function returns the number of data values in the data object **wind**, the cut-point for the new histogram is identified and the new data object

windshift is initialized. In the **R** form of a do loop which follows, 2π is subtracted from any data values that are greater than or equal to $2\pi - 5\pi/8$. This last operation is effectively what shifts the right-hand part of the original histogram to the left of the part beginning at 0. The linear histogram in the right-hand panel of Fig. 2.5 is then obtained using the command:

```
hist(windshift, main=" ", xlab="Wind direction (radians)", ylab="Frequency", breaks=
seq(from=-5*pi/8, to=2*pi-5*pi/8, by=pi/8), col="grey", xlim=c(-5*pi/8, 2*pi-5*pi/8))
```

In this second histogram, negative values along the horizontal axis simply represent directions measured counterclockwise from north and are obviously equivalent to their original positive values (mod 2π). The main features of unimodality and skewness of the data distribution clockwise from north to south are reflected clearly in this new histogram. However, whilst it is far easier to interpret than the first histogram, the reader is still required to equate its two extremes as corresponding to the same point on the unit circle. In the linear histogram the bars to the left are far from those to the right, whilst on the circle they are adjacent. So the problem of representing circular data using a linear device remains.

There are additional modifiers that can be used to further improve the appearance and add information to this type of histogram. As with the construction of any histogram, the best number of class intervals to use will generally need some exploration.

Some authors advocate plotting two repeats of the histogram in order to avoid misleading visual effects stemming from the arbitrary nature of the cut-point. However, we feel that such plots can be open to considerable misinterpretation if the casual viewer misses the key issue that the data are plotted twice (and thus, say, mistakenly interprets unimodal data as being bimodal). Instead, if you want to use a linear histogram, we would recommend plotting only one repeat of the data, but, as we have done, give thought to the most appropriate cut-point to use.

Clearly, the various graphical summaries that we have explored have their advantages and disadvantages. However, as we have advocated, the different circular representations can be combined into composite diagrams with components stressing different levels of detail of the data distribution. In the case of rose diagrams and linear histograms, their visual appearance can be influenced, often strongly, by decisions made about the number of class intervals used to group the data. Sometimes this number is dictated by the grouping used during data collection, otherwise one has considerable flexibility regarding the choice of bin size and thus the appearance of the plot. Bandwidth selection is the analogous issue in kernel density estimation.

3

Circular Summary Statistics

3.1 Introduction

As we saw in the previous chapter, the most natural way of representing a circular observation graphically is as a point on a circle; the circle representing, for example, the face of a compass or a 12 or 24 hour clock, or a means with which to identify the time of the year. The radius of the circle, $r > 0$, is clearly immaterial, and the choice $r = 1$ simplifies things mathematically. For this choice, each observation defines a unit vector, $\mathbf{x}$, directed from the centre of the unit circle to the observation on the unit circle.

Equivalently, once an origin has been chosen (e.g. north, or the one used in mathematics corresponding to the positive horizontal axis) together with an orientation (e.g. clockwise from north or counterclockwise from the mathematical origin), a circular observation can be represented by the angle, θ, subtended by the arc around the unit circle, in the chosen orientation, from the origin to the observation. Negative values of θ correspond to angles measured in the opposite direction to the chosen orientation.

In order to standardize the presentation, henceforth we will generally refer to angles as measured in radians from the positive horizontal axis in a counterclockwise direction. Clearly, the angle used to identify a point on the unit circle is not unique as θ and $\theta + 2\pi p, p = \pm 1, \pm 2, \ldots$, correspond to the same point on the unit circle. Thus, when referring to an angle we will implicitly mean its value modulo 2π. For this choice of units, origin and orientation, the unit vector $\mathbf{x}$ and the angle θ are related through $\mathbf{x} = (\cos\theta, \sin\theta)^T$, the T superscript denoting transpose.

Further mathematical simplification is achieved by representing the unit vector $\mathbf{x}$ in the complex plane rather than in the real plane, the horizontal axis representing its real component and the vertical axis its imaginary component. Using this alternative representation, a circular observation with unit vector $\mathbf{x}$ can be represented as the complex number $z = e^{i\theta} = \cos\theta + i\sin\theta$, i denoting $\sqrt{-1}$. Such a representation of a circular observation is portrayed in Fig. 3.1.

In this chapter we consider those statistical summaries routinely used to summarize a sample of circular data and how they can be calculated using **R**. We start by considering the

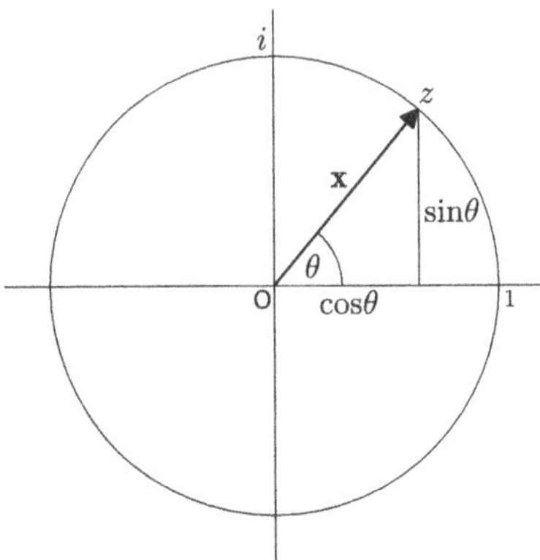

Figure 3.1 A circular observation with unit vector $\mathbf{x}$ represented in the complex plane by the complex number $z = \cos\theta + i\sin\theta$. The arrow identifies the unit vector $\mathbf{x}$ with corresponding angle θ measured counterclockwise from the origin at $z = 1$

sample trigonometric moments, as these provide the basis for most of the measures of location, concentration, dispersion, skewness and kurtosis considered in subsequent sections. The chapter ends with a discussion of the corrections that should be applied when analysing grouped data.

3.2 Sample Trigonometric Moments

Consider a random sample of circular observations of size n, with associated unit vectors $\mathbf{x}_1, \ldots, \mathbf{x}_n$, angles $\theta_1, \ldots, \theta_n$, and complex numbers $z_1, \ldots, z_n$.

For $p = 0, \pm 1, \pm 2, \ldots$ the *pth sample trigonometric moment about the zero direction* is given by

$$t_{p,0} = \frac{1}{n}\sum_{j=1}^{n} z_j^p = \frac{1}{n}\sum_{j=1}^{n} e^{ip\theta_j} = \frac{1}{n}\sum_{j=1}^{n}(\cos p\theta_j + i\sin p\theta_j) = a_p + ib_p, \tag{3.1}$$

where

$$a_p = \frac{1}{n}\sum_{j=1}^{n}\cos p\theta_j, \quad b_p = \frac{1}{n}\sum_{j=1}^{n}\sin p\theta_j, \tag{3.2}$$

$a_{-p} = a_p$ and $b_{-p} = -b_p$. Clearly, $t_{0,0} = 1$.

The complex number $t_{p,0}$ defines a *mean resultant vector* in the complex plane, of length

$$\bar{R}_p = (a_p^2 + b_p^2)^{1/2} \in [0, 1], \tag{3.3}$$

and with direction

$$\bar{\theta}_p = \mathrm{atan2}(b_p, a_p), \tag{3.4}$$

where

$$\operatorname{atan2}(b_p, a_p) = \begin{cases} \arctan(b_p/a_p), & a_p > 0, \\ \arctan(b_p/a_p) + \pi, & b_p \geq 0, a_p < 0, \\ \arctan(b_p/a_p) - \pi, & b_p < 0, a_p < 0, \\ \pi/2, & b_p > 0, a_p = 0, \\ -\pi/2, & b_p < 0, a_p = 0, \\ \text{undefined}, & b_p = 0, a_p = 0, \end{cases} \tag{3.5}$$

the inverse tangent function, **arctan**, returning values in $(-\pi/2, \pi/2)$. The **atan2** function returns values in $(-\pi, \pi]$ which can be mapped to $[0, 2\pi)$ by adding 2π to any negative values.

When $\bar{R}_p > 0$, the polar representation of $t_{p,0}$ is

$$t_{p,0} = \bar{R}_p e^{i\bar{\theta}_p} = \bar{R}_p(\cos\bar{\theta}_p + i\sin\bar{\theta}_p), \tag{3.6}$$

with $\bar{R}_p$ and $\bar{\theta}_p$ being the *mean resultant length* and *mean direction*, respectively, of the mean resultant vector of $p\theta_1, \ldots, p\theta_n$. It follows from (3.1) and (3.6) that, if $\bar{R}_p > 0$,

$$a_p = \bar{R}_p \cos\bar{\theta}_p, \quad b_p = \bar{R}_p \sin\bar{\theta}_p. \tag{3.7}$$

Throughout this chapter we will illustrate the ideas and summaries introduced using a sample of 22 resultant directions, measured in degrees, moved by 22 Sardinian sea stars over a period of 11 days after displacement from their natural habitat. Pabst and Vicentini (1978) describe the experiment in which the data were collected, whilst Upton and Fingleton (1989, pages 274–5) and Fisher (1993, pages 86–7, 245) provide the data and analyses of them. All but one of the data values are included in the (non-circular) data object **fisherB11** available within **R**'s **circular** package. The missing value corresponds to a resultant direction of 8°. To form the complete sample and two different circular representations of it we use the commands:

```
library(circular)
fB11 <- c(fisherB11, 8)
cfB11 <- circular(fB11, units="degrees", zero=circular(pi/2), rotation="clock")
fB11c <- circular(fB11, units="degrees", zero=circular(0), rotation="counter")
```

The circular data object **cfB11** will be used in the production of Fig. 3.2, and **fB11c** to calculate the values of summary statistics.

Figure 3.2 displays three mean resultant vectors for, from left to right, the data in **cfB11** and their doubled and tripled values (modulo 360°). The lengths of the three mean resultant vectors represent $\bar{R}_1$, $\bar{R}_2$ and $\bar{R}_3$, and their directions $\bar{\theta}_1$, $\bar{\theta}_2$ and $\bar{\theta}_3$, respectively. The latter change only slightly across the three plots but there is a pronounced decrease in the mean resultant lengths consistent with the increasing dispersion across the plots. Movement towards the shore can be equated with the zero direction, whilst 180° corresponds to movement towards the sea.

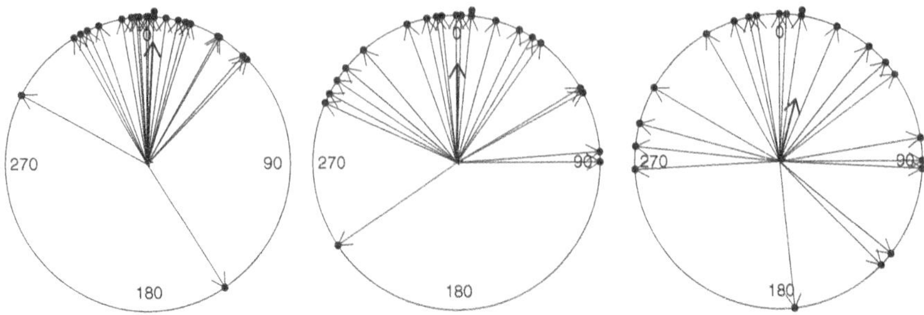

Figure 3.2 Circular plots of the resultant directions of 22 Sardinian sea stars, $\theta_1, \ldots, \theta_{22}$ (left), $2\theta_1, \ldots, 2\theta_{22}$ (centre) and $3\theta_1, \ldots, 3\theta_{22}$ (right), together with arrows representing their unit vectors and bold arrows identifying their mean resultant vectors

The sample mean direction and sample mean resultant length of $\theta_1, \ldots, \theta_n$, $\bar{\theta}_1$ and $\bar{R}_1$, are usually denoted simply by $\bar{\theta}$ and $\bar{R}$, and are referred to unequivocally as *the sample mean direction* and *the sample mean resultant length*, respectively. It follows from (3.2) and (3.7) that, when $\bar{R} > 0$,

$$\frac{1}{n}\sum_{j=1}^{n} \sin(\theta_j - \bar{\theta}) = 0, \quad \frac{1}{n}\sum_{j=1}^{n} \cos(\theta_j - \bar{\theta}) = \bar{R}. \tag{3.8}$$

The circular plot on the left of Fig. 3.2 was produced using the following **R** commands:

```
plot(cfB11, stack=TRUE, bins=720, cex=1.5) ; arrows.circular(cfB11)
arrows.circular(mean(cfB11), y=rho.circular(cfB11), lwd=3)
```

Note the use of the commands **mean(cfB11)** and **rho.circular(cfB11)** to calculate $\bar{\theta}$ and $\bar{R}$ for the data in **cfB11**. Alternatively, the values of $\bar{\theta}$ and $\bar{R}$, as well as those of a_1 and b_1, can be computed using the commands:

```
t10 <- trigonometric.moment(fB11c, p=1)
tbar <- t10$mu ; Rbar <- t10$rho ; a1 <- t10$cos ; b1 <- t10$sin
tbar ; Rbar ; a1 ; b1
```

From the output produced, $\bar{\theta} = 3.10°$, $\bar{R} = 0.83$, $a_1 = 0.83$ and $b_1 = 0.04$. Similarly, using the commands:

```
t20 <- trigonometric.moment(fB11c, p=2)
tbar2 <- t20$mu ; Rbar2 <- t20$rho ; a2 <- t20$cos ; b2 <- t20$sin
tbar2 ; Rbar2 ; a2 ; b2
```

we obtain $\bar{\theta}_2 = 0.64°$, $\bar{R}_2 = 0.67$, $a_2 = 0.67$ and $b_2 = 0.01$.

Sample trigonometric moments about the mean direction $\bar{\theta}$ can also be defined and prove useful. The *pth sample trigonometric moment about the mean direction* is given by

$$t_{p,\bar{\theta}} = \frac{1}{n}\sum_{j=1}^{n} e^{ip(\theta_j-\bar{\theta})} = \frac{1}{n}\sum_{j=1}^{n}\{\cos p(\theta_j-\bar{\theta}) + i\sin p(\theta_j-\bar{\theta})\} = \bar{a}_p + i\bar{b}_p, \tag{3.9}$$

where

$$\bar{a}_p = \frac{1}{n}\sum_{j=1}^{n}\cos p(\theta_j-\bar{\theta}), \quad \bar{b}_p = \frac{1}{n}\sum_{j=1}^{n}\sin p(\theta_j-\bar{\theta}). \tag{3.10}$$

From (3.8),

$$\bar{a}_1 = \bar{R}, \quad \bar{b}_1 = 0, \tag{3.11}$$

and thus $t_{1,\bar{\theta}} = \bar{R}$. Consequently, for the centred sample, $\theta_1 - \bar{\theta}, \ldots, \theta_n - \bar{\theta}$, the mean direction is 0 but the mean resultant length is unchanged.

The components of the second sample trigonometric moment about the mean direction for the sea star data can be computed using the commands:

```
t2t <- trigonometric.moment(fB11c, p=2, center=TRUE)
abar2 <- t2t$cos ; bbar2 <- t2t$sin ; abar2 ; bbar2
```

which output $\bar{a}_2 = 0.67$ and $\bar{b}_2 = -0.06$. The interpretation of $\bar{a}_2$ and $\bar{b}_2$ will be considered in Section 3.5.

As we will see in the following sections, most of the measures used to summarize circular data are related to the sample trigonometric moments. The population analogues of the sample trigonometric moments will be introduced in Section 4.2.2 and play an important role in circular distribution theory.

3.3 Measures of Location

3.3.1 *Sample Mean Direction*

The sample mean direction, $\bar{\theta} \equiv \bar{\theta}_1$, defined in Section 3.2, is the most commonly used measure of location for circular data. When it exists, it corresponds to the direction of the mean resultant vector of the data. It will not exist, for instance, for any sample of even size of the form $\theta_1, \ldots, \theta_{n/2}, \theta_1 + \pi, \ldots, \theta_{n/2} + \pi$; the obvious pairs of unit vectors cancelling each other out and leading to $\bar{R} = 0$. Other forms of cyclic samples, for which triples, quadruples, etc. of their unit vectors cancel each other out, and thus also have $\bar{R} = 0$, can arise too. For unimodal samples that are close to symmetric, $\bar{\theta}$ provides a good measure of central location.

The mean direction can be shown to be equivariant under rotation; i.e. if a sample of circular data with mean direction $\bar{\theta}$ is rotated clockwise through an angle ψ, the mean direction of the rotated sample will be $\bar{\theta} + \psi \pmod{2\pi}$.

In the previous section, the mean direction for the sea star data was found to be $\bar{\theta} = 3.10°$, which indicates that over the 11 days after displacement the mean direction of movement of the 22 sea stars was almost directly inshore. This can be calculated using either

```
mean(fB11c)
```

or

```
t10 <- trigonometric.moment(fB11c, p=1) ; tbar <- t10$mu
```

3.3.2 *Sample Median Direction*

The *sample median direction*, denoted by $\tilde{\theta}$, provides a robust alternative to the sample mean direction, $\bar{\theta}$, as a measure of location. It is particularly useful when the sample is skewed. A sample median direction is defined as any angle ψ for which half of the data points lie in the arc $[\psi, \psi + \pi)$ and the majority of the points are nearer to ψ than to $\psi + \pi$. From this definition, it is clear that the median direction need not be unique. When n is odd, a median direction will correspond to one of the data points. When n is even, it is usually taken to be the mean of those data points immediately to its left and right. Formally, a median direction can be identified by minimizing the dispersion measure (see Section 3.4.3)

$$d_2(\psi) = \frac{1}{n}\sum_{j=1}^{n}\{\pi - |\pi - |\theta_j - \psi||\}, \qquad (3.12)$$

and this is the approach implemented using the **circular** package's **medianCircular** function. Note however that this function will identify just one median direction, not all median directions should more than one exist, and so the results obtained from it should be treated with caution and interpreted with the aid of graphical representation of the data.

For the sea star data, the median direction is obtained using the command:

```
medianCircular(fB11c)
```

which returns a value of 1.54°, corresponding to a direction close to directly inshore once more. For these data, then, there is little difference between the mean and median directions. This similarity between their values is consistent with the distribution of the data being close to symmetric.

3.4 Measures of Concentration and Dispersion

3.4.1 *Sample Mean Resultant Length*

The sample mean resultant length, $\bar{R} \equiv \bar{R}_1 \in [0, 1]$, defined in Section 3.2, is the most commonly used measure of the *concentration* of unimodal circular data. $\bar{R}$ equals 1 only when all the data points are located at the same point on the unit circle (or, equivalently, all the unit vectors are identical). A value close to 1 indicates that the data are closely clustered around the mean direction. When the data are spread evenly around the circle, $\bar{R}$ will take a value near 0. However, a value of 0 for $\bar{R}$ should not be interpreted as necessarily indicating that the data are evenly distributed around the circle because, as we saw in Section 3.3.1, $\bar{R} = 0$ for any sample with a cyclic structure.

In Section 3.2, the mean resultant length for the sea star data was found to be 0.83, reflecting the fact that all but two of the 22 resultant directions are fairly concentrated around the mean direction. That mean resultant length can be calculated using either

```
Rbar <- rho.circular(fB11c)
```

or

```
t10 <- trigonometric.moment(fB11c, p=1) ; Rbar <- t10$rho
```

3.4.2 *Sample Circular Variance and Standard Deviation*

The *sample circular variance* is defined as

$$V = 1 - \bar{R}. \tag{3.13}$$

Since $\bar{R} \in [0, 1]$, $V \in [0, 1]$ too. Batschelet (1981, page 34) uses the term 'angular variance' to refer to $2(1 - \bar{R}) \in [0, 2]$, a measure of circular variability calculated using the **angular.variance** function available within the **circular** package.

The *sample circular standard deviation* is given by

$$\hat{\sigma} = \{-2 \log(1 - V)\}^{1/2} = \{-2 \log \bar{R}\}^{1/2} \subset [0, \infty], \tag{3.14}$$

and can be computed in the **circular** library using the **sd.circular** function. For concentrated samples, with small V,

$$\hat{\sigma} \simeq (2V)^{1/2} = \{2(1 - \bar{R})\}^{1/2}, \tag{3.15}$$

which explains the definition of the angular variance as an alternative to the circular variance (3.13). Batschelet (1981, page 34) refers to $\{2(1 - \bar{R})\}^{1/2}$ as the (*mean*) *angular deviation*. It can be calculated using the **angular.deviation** function within the **circular** package.

For the sea star data, the values of V and $\hat{\sigma}$ can be computed using the following commands which incorporate the previously calculated object **Rbar**:

```
V <- 1-Rbar ; V ; sd.circular(fB11c)
```

The values returned are $V = 0.17$ and $\hat{\sigma} = 0.61$. Both measures of dispersion are relatively low, reflecting the fact that the sea star data are fairly concentrated.

Of the two measures, we have a preference for V, as the $[0, 1]$ scale appears more natural than $[0, 2]$. However, for whichever measure you choose, make sure that you are consistent in its use, and be clear in your text as to which measure you are using: clearly *variance* is an ambiguous term. In interpreting published works, detective work may be required: are values ever greater than one, how exactly is the measure described, and is Batschelet cited in support of the measure used?

3.4.3 Other Sample Dispersion Measures

The so-called *sample circular dispersion* is defined as

$$\hat{\delta} = \frac{1 - \bar{R}_2}{2\bar{R}^2}, \tag{3.16}$$

$\bar{R}_2$ being the mean resultant length of the doubled angles, $2\theta_1, \ldots, 2\theta_n$, as defined in (3.3). Using the **R** objects computed previously in Section 3.2, its value for the sea star data can be obtained using the commands:

```
delhat <- (1-Rbar2)/(2*Rbar**2) ; delhat
```

which return $\hat{\delta} = 0.24$.

There are two commonly used *measures of the distance* between two angles, ψ and ω, which lead to two alternative measures of dispersion for circular data. The first distance measure is

$$1 - \cos(\psi - \omega), \tag{3.17}$$

with the associated measure of dispersion of $\theta_1, \ldots, \theta_n$ about ψ,

$$d_1(\psi) = \frac{1}{n}\sum_{j=1}^{n}\{1 - \cos(\theta_j - \psi)\}. \tag{3.18}$$

The dispersion measure $d_1(\psi)$ is minimized when $\psi = \bar{\theta}$, with $d_1(\bar{\theta}) = 1 - a_1 = 1 - \bar{R} = V$. Using (3.17) as the distance measure, the mean distance between data points is

$$\bar{d}_1 = \frac{1}{n^2}\sum_{j=1}^{n}\sum_{k=1}^{n}\{1 - \cos(\theta_j - \theta_k)\} = 1 - \bar{R}^2.$$

The second commonly used distance measure is

$$\min(\psi - \omega, 2\pi - (\psi - \omega)) = \pi - |\pi - |\psi - \omega||, \tag{3.19}$$

with its associated dispersion measure (3.12). As explained in Section 3.3.2, $d_2(\psi)$ is minimized when $\psi = \tilde{\theta}$, $d_2(\tilde{\theta})$ being referred to as the *circular mean deviation*. Using (3.19) as the distance measure, the mean distance between data points is

$$\bar{d}_2 = \frac{1}{n^2}\sum_{j=1}^{n}\sum_{k=1}^{n}\{\pi - |\pi - |\theta_j - \theta_k||\} \in [0, \pi/2].$$

The values of $\bar{d}_1$ and $d_2(\tilde{\theta})$ for the sea star data can be computed using the commands:

```
dbar1 <- 1-(Rbar**2) ; dbar1
mc <- medianCircular(fB11c, deviation=TRUE) ; mc$deviation
```

which output the values $\bar{d}_1 = 0.31$ and $d_2(\tilde{\theta}) = 0.43$.

Finally, the *circular range*, *w*, is defined as the length of the smallest arc containing all of the observations. It can be computed using the **circular** package's **range.circular** function. Using the command:

```
range.circular(fB11c)
```

produces $w = 209°$ for the sea star data. This relatively large circular range is primarily due to the rather atypical resultant directions of two of the sea stars noted earlier. We can explore the effect of the removal of these two data points on some of the summary statistics using the commands:

```
fB11cred <- fB11c[-c(12,13)]
mean(fB11cred) ; medianCircular(fB11cred, deviation=TRUE)
rho.circular(fB11cred) ; range.circular(fB11cred)
```

For the reduced data set in **fB11cred, R** computes $\bar{\theta} = 4.08°$, $\tilde{\theta} = 1.88°$, $d_2(\tilde{\theta}) = 0.29$, $\bar{R} = 0.93$ and $w = 76°$. Comparing these values with those for the complete data set, we see that, as might have been expected, $\bar{\theta}$ and $\tilde{\theta}$ have changed only very slightly. On the other hand, the concentration and dispersion measures have changed substantially: $\bar{R}$ increasing by 0.1, and $d_2(\tilde{\theta})$ and w decreasing by 0.14 and a massive 133°, respectively.

We provide the details of these alternative measures primarily for reference, as you will encounter their use in the literature. Our recommendation is to use the popular mean resultant length ($\bar{R}$) or circular variance (V) to summarize the spread of unimodal samples. For multimodal samples the range (or perhaps an analogous inter-quartile range) may be the most robust and easily interpreted measure.

3.5 Measures of Skewness and Kurtosis

As we saw in (3.11), the elements of the first trigonometric moment about the mean direction, $t_{1,\bar{\theta}}$, are $\bar{a}_1 = \bar{R}$ and $\bar{b}_1 = 0$. Thus $\bar{a}_1$ is a measure of concentration and $\bar{b}_1$ is not useful as a summary statistic. The elements of the second trigonometric moment about the mean direction, $t_{2,\bar{\theta}}$, provide the basic measures of skewness and kurtosis for circular data.

The second central sine moment

$$\bar{b}_2 = \frac{1}{n}\sum_{j=1}^{n} \sin 2(\theta_j - \bar{\theta}) = \bar{R}_2 \sin(\bar{\theta}_2 - 2\bar{\theta}), \tag{3.20}$$

is the first non-zero central sine moment and provides a measure of the skewness of the data points around the mean direction. For unimodal data sets, $\bar{b}_2$ will be close to zero if the data distribution is near symmetric about $\bar{\theta}$, and relatively large and *negative* (*positive*) when the distribution of the data is skewed in the *counterclockwise* (*clockwise*) direction away from the mean direction.

The second central cosine moment

$$\bar{a}_2 = \frac{1}{n}\sum_{j=1}^{n}\cos 2(\theta_j - \bar{\theta}) = \bar{R}_2\cos(\bar{\theta}_2 - 2\bar{\theta}), \quad (3.21)$$

is a measure of the kurtosis of circular data. It takes the value 1 when the data points are all identical, corresponding to maximum peakedness, and values close to 0 for data near evenly distributed around the circle.

The properties of $\bar{b}_2$ and $\bar{a}_2$ for concentrated circular distributions led Mardia (1972, Section 2.7.2) to propose the standardized measures of skewness and kurtosis

$$\hat{s} = \frac{\bar{b}_2}{(1-\bar{R})^{3/2}} \quad (3.22)$$

and

$$\hat{k} = \frac{\bar{a}_2 - \bar{R}^4}{(1-\bar{R})^2}. \quad (3.23)$$

The interpretation of $\hat{s}$ is similar to that of $\bar{b}_2$. For data from a wrapped normal distribution (see Section 4.3.7), $\bar{a}_2 - \bar{R}^4$ and $\hat{k}$ will be close to zero. As $\bar{R} \in [0, 1]$, the magnitudes of $\hat{s}$ and $\hat{k}$ will generally be larger than those of their unstandardized counterparts, $\bar{b}_2$ and $\bar{a}_2 - \bar{R}^4$.

Making use of the **R** objects previously computed in this chapter, the values of $\hat{s}$ and $\hat{k}$ can be calculated for the sea star data using the commands:

```
hats <- bbar2/(V**(3/2)) ; hats
hatk <- (abar2-Rbar**4)/(V**2) ; hatk
```

R returns the values $\hat{s} = -0.92$ and $\hat{k} = 6.64$. As we saw in Section 3.2, for the same data, $\bar{b}_2 = -0.06$, $\bar{a}_2 = 0.67$ and $\bar{R} = 0.83$; so $\bar{a}_2 - \bar{R}^4 = 0.20$. The values of $\bar{b}_2$ and $\hat{s}$ both indicate that the sea star data are somewhat skewed in the counterclockwise direction, whilst those for $\bar{a}_2 - \bar{R}^4$ and $\hat{k}$ suggest that the data are more peaked than one might expect of a sample of the same size from a wrapped normal distribution with the same value of $\bar{R}$. However, it is not obvious whether their departures from 0 are significant or not. In Sections 4.2.5, 5.2 and 5.3 we will consider techniques which can be used to draw inferences about the population analogues of these statistical summaries.

3.6 Corrections for Grouped Data

In many contexts in which circular data are recorded it can prove difficult to measure angles precisely. Under such circumstances, the circumference of the unit circle is generally divided into an even number of sectors, with arcs of equal length, and the frequencies of data points falling in the different sectors are what is actually recorded. The resulting data are clearly grouped, the grouping being more or less coarse depending on the arc length employed.

When it comes to analysing such frequency data, summary statistics are usually calculated with the data points falling in an arc all placed at the centre of the arc. In general, the values of any mean resultant directions, the $\bar{\theta}_p$, calculated in this way will differ only slightly from the values that would have been obtained had the data been measured precisely. However, the values of any mean resultant lengths, the $\bar{R}_p$, will generally be somewhat smaller than they would have been. This is important because, as we have seen, $\bar{R}$ and $\bar{R}_2$ appear in the definitions of some of the statistics most commonly used to summarize circular data.

The correction advocated by Mardia (1972, Section 2.7.2) is to calculate instead

$$\bar{R}_p^* = c(p\psi)\bar{R}_p, \tag{3.24}$$

where the correction term is given by

$$c(p\psi) = \frac{p\psi/2}{\sin(p\psi/2)},$$

and ψ is the angle, measured in radians, subtended by each arc. For example, for 16 arcs of equal length, $\psi = \pi/8$. The magnitude of the correction factor $c(p\psi)$ increases with increasing p and the coarseness of the grouping as represented by ψ.

To illustrate the effects of grouping and the use of correction for it, suppose the 22 sea star resultant directions had been recorded not to the nearest degree but in terms of their falling in one of 16 or 8 arcs of equal length, the first arc starting at the zero direction. The two arc-centred data sets corresponding to these two groupings can be set up in **R** using the commands:

```
fB11g1 <- rep(c(11.25, 33.75, 146.25, 303.75, 326.25, 348.75), c(8, 4, 1, 1, 3, 5))
fB11g1c <- circular(fB11g1,units="degrees", zero=circular(0), rotation="counter")
fB11g2 <- rep(c(22.5, 157.5, 292.5, 337.5), c(12, 1, 1, 8))
fB11g2c <- circular(fB11g2,units="degrees", zero=circular(0), rotation="counter")
```

To calculate the values of $\bar{\theta}$, $\bar{R}$, $\bar{\theta}_2$ and $\bar{R}_2$, and the corrected measures $\bar{R}^* \equiv \bar{R}_1^*$ and $\bar{R}_2^*$, for the data grouped using 16 arcs, we can run the commands:

```
t10 <- trigonometric.moment(fB11g1c, p=1)
tbar <- t10$mu ; Rbar <- t10$rho ; tbar ; Rbar
t20 <- trigonometric.moment(fB11g1c, p=2)
tbar2 <- t20$mu ; Rbar2 <- t20$rho ; tbar2 ; Rbar2
corr1 <- ((2*pi/16)/2)/sin((2*pi/16)/2)
corr2 <- ((2*2*pi/16)/2)/sin((2*2*pi/16)/2)
Rbars <- corr1*Rbar ; Rbar2s <- corr2*Rbar2 ; Rbars ; Rbar2s
```

which return the values $\bar{\theta} = 2.70°$, $\bar{R} = 0.83$, $\bar{\theta}_2 = 0.87°$, $\bar{R}_2 = 0.67$, $\bar{R}^* = 0.84$ and $\bar{R}_2^* = 0.69$. These compare with the values $\bar{\theta} = 3.10°$, $\bar{R} = 0.83$, $\bar{\theta}_2 = 0.64°$ and $\bar{R}_2 = 0.67$ for the original ungrouped data. Thus, for this grouping of the data, the first two mean directions indeed change only very slightly and there is in fact no need to apply any correction to the first two mean resultant lengths.

Similarly, for the data obtained using the coarser grouping, we can use the commands:

```
t10 <- trigonometric.moment(fB11g2c, p=1)
tbar <- t10$mu ; Rbar <- t10$rho ; tbar ; Rbar
t20 <- trigonometric.moment(fB11g2c, p=2)
tbar2 <- t20$mu ; Rbar2 <- t20$rho ; tbar2 ; Rbar2
corr1 <- ((2*pi/8)/2)/sin((2*pi/8)/2)
corr2 <- ((2*2*pi/8)/2)/sin((2*2*pi/8)/2)
Rbars <- corr1*Rbar ; Rbar2s <- corr2*Rbar2 ; Rbars ; Rbar2s
```

which output $\bar{\theta} = 3.15°$, $\bar{R} = 0.82$, $\bar{\theta}_2 = 5.71°$, $\bar{R}_2 = 0.65$, $\bar{R}^* = 0.84$ and $\bar{R}_2^* = 0.72$. Now the change in $\bar{\theta}_2$ is more pronounced. The correction to the mean resultant lengths needlessly inflates their values, especially that obtained using $\bar{R}_2^*$.

Hence, our advice, following Jammalamadaka and SenGupta (2001, page 22), is only to apply this correction if the data are divided into less than eight segments. At least when conducting accurate scientific work, such situations will rarely be encountered in practice.

3.7 Axial Data

Particularly in geology, but also in various other scientific disciplines, the orientations of axes are often of interest. For instance, when studying the orientations of the stones, pebbles or rocks deposited by a receding glacier, a glaciologist might measure the orientation of the principal axis of each piece of undisturbed debris. The resulting axial data differ from circular data in the sense that the two angles θ and $\theta + \pi$ (or $\theta + 180°$) define the same axial orientation.

An example of axial data is provided by the object **fisherB2c** within **R**'s **circular** library. The data, presented in Fisher (1993, page 240), represent the long-axis orientations of 133 feldspar laths in basalt rock recorded between 0° and 180°. A circular plot portraying the original data, $\theta_1, \ldots, \theta_{133}$, using solid black circles, and their antipodal values, $\theta_1 + 180°, \ldots, \theta_{133} + 180°$, using grey circles, is provided in the left-hand panel of Fig. 3.3. A line drawn to connect a solid black circle to its antipodal grey circle represents the long-axis orientation of one of the feldspar laths.

The standard approach to analysing axial data is to double their angles, so as to obtain circular data, calculate any required summary statistics for the doubled angles and transform back, in order to obtain the mean or median axis, by dividing the mean or median direction by 2. The recommendation of Fisher (1993, page 37) is that any measures of concentration or spread should be quoted for the doubled angles and not be back-transformed. The doubled angles for the long-axis orientations are shown in the right-hand panel of Fig. 3.3. The mean and median directions of the doubled angles, together with their mean resultant length, can be obtained in **R** by running the commands:

```
data(fisherB2c) ; fB2cdouble <- 2*fisherB2c
mean(fB2cdouble) ; medianCircular(fB2cdouble) ; rho.circular(fB2cdouble)
```

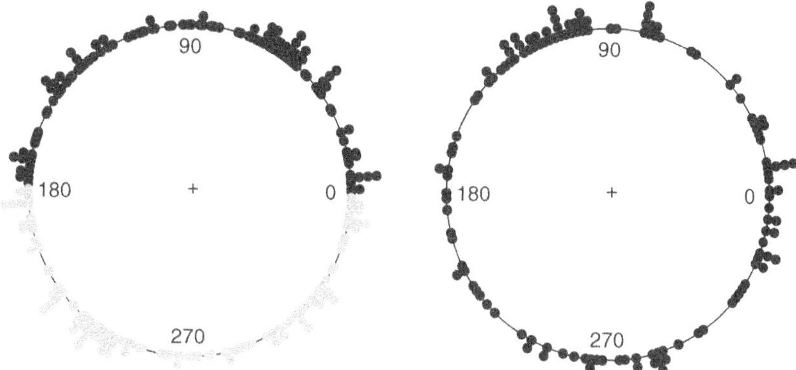

Figure 3.3 Circular data plots of: (left) the original long-axis orientations of 133 feldspar laths (black circles) and their antipodal values (grey circles); (right) the doubled angles of the long-axis orientations

which produce $\bar{\theta} = 143.8°$, $\tilde{\theta} = 204.0°$ and $\bar{R} = 0.12$. The mean axis is therefore $143.8°/2 = 71.9°$, and the median axis $204.0°/2 = 102°$. The disparity between the two reflects the fact that, as can be seen clearly in the right-hand panel of Fig. 3.3, the data distribution is multimodal. As measures of concentration and spread one could quote the values of $\bar{R} = 0.12$ and $V = 1 - \bar{R} = 0.88$ for the doubled angles.

4

Distribution Theory and Models for Circular Random Variables

4.1 Introduction

In this chapter we consider basic distribution theory and models for circular random variables. As we shall see, the properties of circular distribution functions and probability densities have much in common with those of their linear counterparts. However, there are also important differences that result from the need to specify the two circular functions uniquely and, at the same time, reflect the periodicity of circular distributions. The population analogues of the trigonometric-moment-based measures introduced in Chapter 3 are functions of the Fourier coefficients, or complex numbers making up the characteristic function, of a circular distribution. We consider classical models for circular data as well as some of the more flexible models proposed recently in the literature.

4.2 Circular Distribution Theory

As we saw in Chapter 3, there are various ways of representing a circular observation. Nevertheless, it is usual to specify a circular distribution as being that of a random angle, which we will denote by Θ, corresponding to a point on the circumference of the unit circle. This random angle is the population analogue of the angular observations referred to in Chapter 3. As there, when defining a random angle it is necessary to choose an initial direction (the 'zero' direction), an orientation (clockwise or counterclockwise) and the units of measurement. In an attempt to keep things simple we will generally use the 'mathematical' representation of an angle throughout this chapter, with the zero direction corresponding to the positive horizontal axis, and angles measured in an counterclockwise direction in radians.

Most models for circular distributions assume Θ to be (absolutely) continuous and can be specified through a probability density function. Discrete models, specified through a (countable) set of probabilities, are also available but their use is less common. Models with discrete as well as continuous components may be relevant in applications.

4.2.1 Circular Distribution and Probability Density Functions

By analogy with the approach used for random variables observed on the real line, a general means of specifying a circular probability distribution is via its distribution function. Nevertheless, there are two important complications associated with the specification of a circular distribution function, which arise because of the periodicity of circular random variables. The first is that the definition of a circular distribution function depends on the choice of initial direction, orientation and units of measurement used. Secondly, the fact that the angles θ and $\theta + 2k\pi$, $k = 0, \pm 1, \pm 2, \ldots$, correspond to the same point on the unit circle implies that, when specified for any θ-value on the whole of the real line, a circular distribution should be *periodic*. So, for example, the probability (if the distribution is discrete) or the density (if the distribution is absolutely continuous) should be the same for the angles 30° and 750°, as both correspond to the same point on the unit circle. It is thus usual to specify a circular distribution function, F, as the function given by

$$F(\theta) = P(0 < \Theta \leq \theta), \quad 0 \leq \theta \leq 2\pi, \tag{4.1}$$

and

$$F(\theta + 2\pi) - F(\theta) = 1, \quad -\infty < \theta < \infty. \tag{4.2}$$

Equation (4.1) is a circular analogue of the usual definition of a distribution function for a random variable observed on the real line. Note in its definition, however, the role of the zero direction and the restriction to θ-values between 0 and 2π. Equation (4.2) is an extra condition imposed to reflect the periodicity of a circular distribution. This property is related to, but is clearly not the direct analogue of, the property that the total probability associated with a linear distribution should be 1. It can be interpreted as stating that the probability of obtaining a point on the unit circle within any arc of length 2π radians is 1. This important difference with respect to the distribution function of a linear random variable implies that

$$\lim_{\theta \to -\infty} F(\theta) = -\infty, \quad \lim_{\theta \to \infty} F(\theta) = \infty.$$

In general, then, the values taken by F are clearly not probabilities. However, for $\phi \leq \psi \leq \phi + 2\pi$,

$$P(\phi < \Theta \leq \psi) = F(\psi) - F(\phi) = \int_{\phi}^{\psi} dF(\theta), \tag{4.3}$$

where the integral is a Lebesgue–Stieltjes integral. By definition,

$$F(0) = 0, \quad F(2\pi) = 1.$$

As usual, the distribution function is right-continuous.

An important related function is the *quantile function*, which for $0 \leq u \leq 1$ is defined as

$$Q(u) = \inf\{\theta : F(\theta) \geq u\}, \tag{4.4}$$

i.e. the minimum value of θ for which $F(\theta) = P(0 < \Theta \leq \theta)$ is at least u.

If the circular distribution function F is absolutely continuous then it has a circular probability density function, f, such that

$$\int_{\phi}^{\psi} f(\theta)\mathrm{d}\theta = F(\psi) - F(\phi), \quad -\infty < \phi \leq \psi < \infty.$$

Clearly, the value taken by this integral need not be a probability (for instance, if $\psi = \phi + 4\pi$). A function f is the probability density function of an absolutely continuous circular distribution if and only if

1. $f(\theta) \geq 0$ almost everywhere on $(-\infty, \infty)$,
2. $f(\theta + 2\pi) = f(\theta)$ almost everywhere on $(-\infty, \infty)$,
3. $\int_0^{2\pi} f(\theta)\mathrm{d}\theta = 1$.

The first property also holds for the density of a random variable observed on the real line. The other two are consequences of the periodicity of a circular distribution.

Figure 4.1 illustrates the forms of the distribution function, $F(\theta)$, and the density, $f(\theta)$, for a von Mises distribution (see Section 4.3.8) with mean direction $\mu = \pi$ and

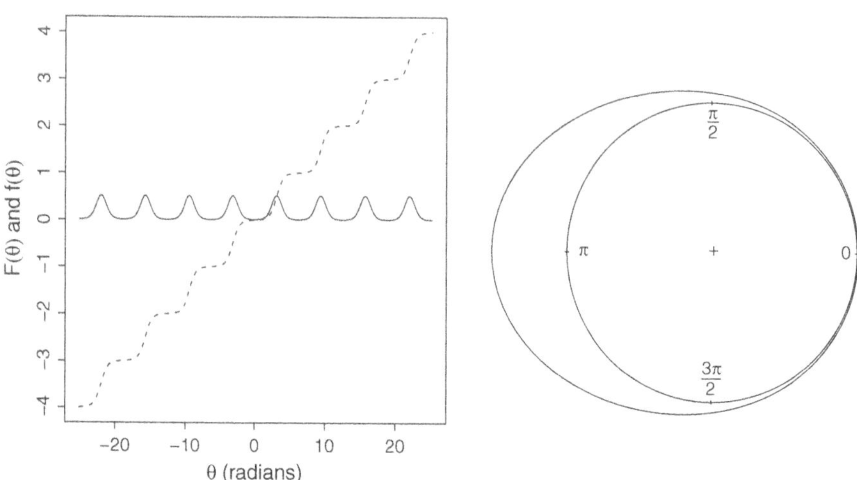

Figure 4.1 Linear representation of the distribution function, $F(\theta)$ (dashed), and density function, $f(\theta)$ (solid), for θ ranging between -8π and 8π radians (left), and polar representation of $f(\theta)$ (right), for a von Mises distribution with mean direction $\mu = \pi$ and concentration parameter $\kappa = 2$

concentration parameter $\kappa = 2$. The linear plot on the left provides a partial representation of both, with θ ranging between -8π and 8π radians. In this linear representation, $f(\theta)$ is multimodal, with the shape of the density repeating itself every 2π radians. In contrast, the polar representation of $f(\theta)$, on the right, is unimodal. As circular densities are periodic, linear representations of them are often just plotted for an interval of θ-values of width 2π radians chosen carefully to ensure an unequivocal interpretation of any modes.

4.2.2 Circular Characteristic Function, Trigonometric Moments and Fourier Series Expansion

In Section 3.1 we saw how useful it is to represent a circular observation as a complex number. By analogy, here we consider the random variable $Z = e^{i\Theta}$ and, related to it, the characteristic function of the random angle Θ. The latter offers an alternative means of describing the distribution of Θ and provides the population analogues of the sample trigonometric moments introduced in Section 3.1.

Consider the function

$$E(Z^t) = E(e^{it\Theta}) = \int_0^{2\pi} e^{it\theta}\, \mathrm{d}F(\theta). \tag{4.5}$$

As Θ is periodic and thus has the same distribution as $\Theta + 2\pi$, the two should have identical values of (4.5); i.e.

$$E(e^{it\Theta}) = E(e^{it(\Theta+2\pi)}) = e^{it2\pi} E(e^{it\Theta}).$$

For the last equality to hold, either $E(e^{it\Theta}) = 0$ or $e^{it2\pi} = 1$. As the former is not useful, we assume the latter to hold—which means that t must be an integer. The characteristic function of Θ is then defined as the doubly infinite sequence of complex numbers $\{\tau_{p,0} : p = 0, \pm 1, \pm 2, \ldots\}$, where

$$\tau_{p,0} = E(e^{ip\Theta}) = \int_0^{2\pi} e^{ip\theta}\, \mathrm{d}F(\theta), \quad p = 0, \pm 1, \pm 2, \ldots \tag{4.6}$$

The $\tau_{p,0}$ are the Fourier coefficients of F. The complex number $\tau_{p,0}$ is referred to as the *pth trigonometric moment of* Θ *about the zero direction*, and can be represented as

$$\tau_{p,0} = \alpha_p + i\beta_p, \tag{4.7}$$

where

$$\alpha_p = E[\cos p\Theta] = \int_0^{2\pi} \cos p\theta\, \mathrm{d}F(\theta) \tag{4.8}$$

and

$$\beta_p = E[\sin p\Theta] = \int_0^{2\pi} \sin p\theta\, \mathrm{d}F(\theta). \tag{4.9}$$

The cosine and sine moments, α_p and β_p, are the population analogues of a_p and b_p defined in Equation (3.2). As the sequence $\{(\alpha_p, \beta_p) : p = 0, \pm 1, \pm 2, \ldots\}$ is equivalent to the characteristic function of Θ, any circular distribution is completely determined by its sine and cosine moments. Clearly,

$$\alpha_{-p} = \alpha_p, \quad |\alpha_p| \in [0, 1], \quad \beta_{-p} = -\beta_p, \quad |\beta_p| \in [0, 1], \tag{4.10}$$

and

$$\tau_{0,0} = 1, \quad \bar{\tau}_{p,0} = \tau_{-p,0}, \quad |\tau_{p,0}| = ||E(e^{ip\Theta})|| \leq E||e^{ip\Theta}|| = 1, \tag{4.11}$$

where $\bar{\tau}_{p,0}$ denotes the complex conjugate of $\tau_{p,0}$ and $||z|| = ||x + iy|| = \sqrt{x^2 + y^2}$ is the norm of the complex number z. Equivalently, then, each $\tau_{p,0}$ can be considered as defining a *mean resultant vector* in the complex plane, of length

$$\rho_p = |\tau_{p,0}| = \{\alpha_p^2 + \beta_p^2\}^{1/2} \in [0, 1], \tag{4.12}$$

and with direction

$$\mu_p = \text{atan2}(\beta_p, \alpha_p), \tag{4.13}$$

where the atan2 function is as defined in Equation (3.5). Equations (4.12) and (4.13) are the population analogues of $\bar{R}_p$ and $\bar{\theta}_p$ defined in Equations (3.3) and (3.4). The special cases ρ_1 and μ_1 are fundamental measures of concentration and location, respectively, and are frequently denoted simply as ρ and μ and referred to unequivocally as *the population mean resultant length* and *the population mean direction*. For the polar representation,

$$\tau_{p,0} = \rho_p e^{i\mu_p} = \rho_p(\cos \mu_p + i \sin \mu_p), \tag{4.14}$$

and hence

$$\alpha_p = \rho_p \cos \mu_p, \quad \beta_p = \rho_p \sin \mu_p. \tag{4.15}$$

Equations (4.14) and (4.15) are the population analogues of Equations (3.6) and (3.7).

If $\sum_{p=1}^{\infty}(\alpha_p^2 + \beta_p^2) = \sum_{p=1}^{\infty} \rho_p^2$ is convergent, Θ has a density $f(\theta)$ which is defined almost everywhere by the Fourier expansion

$$\begin{aligned} f(\theta) = \frac{1}{2\pi} \sum_{p=-\infty}^{\infty} \tau_{p,0}\, e^{-ip\theta} &= \frac{1}{2\pi} \left\{ 1 + 2 \sum_{p=1}^{\infty} (\alpha_p \cos p\theta + \beta_p \sin p\Theta) \right\} \\ &= \frac{1}{2\pi} \left\{ 1 + 2 \sum_{p=1}^{\infty} \rho_p \cos(p\theta - \mu_p) \right\}. \end{aligned} \tag{4.16}$$

By analogy with Equation (3.9), the *pth population trigonometric moment about the mean direction* is defined as

$$\tau_{p,\mu} = E[e^{ip(\Theta-\mu)}] = \bar{\alpha}_p + i\bar{\beta}_p, \tag{4.17}$$

where

$$\bar{\alpha}_p = E[\cos p(\Theta - \mu)], \quad \bar{\beta}_p = E[\sin p(\Theta - \mu)], \tag{4.18}$$

are the population analogues of $\bar{a}_p$ and $\bar{b}_p$ defined in Equation (3.10) and are referred to as the *pth population central cosine and sine moments*. From its definition,

$$\tau_{p,\mu} = \tau_{p,0}\, e^{-ip\mu} = \rho_p[\cos(\mu_p - p\mu) + i\sin(\mu_p - p\mu)]. \tag{4.19}$$

4.2.3 *Basic Population Measures*

Here we consider the population analogues of the sample circular measures introduced in Sections 3.3–3.5. As there, we start with circular measures of location, concentration and dispersion before proceeding to circular measures of skewness and kurtosis.

As we observed in Section 4.2.2, the population mean direction, $\mu = \mu_1$, defined in Equation (4.13), is the basic circular measure of *location*. Importantly, μ is undefined when the mean resultant length, ρ, takes the value 0. A fundamental circular distribution for which $\rho = 0$ is the continuous circular uniform distribution to be studied in Section 4.3.3. When μ does exist for a given distribution, the effect of rotating the distribution through an angle ϕ results in the mean direction changing to $\mu + \phi \pmod{2\pi}$. Thus μ is said to be equivariant under rotation.

An alternative circular measure of location is the *population median direction*, $\tilde{\mu}$, the analogue of the sample median direction, $\tilde{\theta}$, introduced in Section 3.3.2. Formally, $\tilde{\mu}$ is defined as any angle ψ which minimizes

$$E\left[\pi - |\pi - |\Theta - \psi||\right]. \tag{4.20}$$

The median direction need not be unique, although it will be for any circular distribution whose polar representation is unimodal. Any median direction satisfies

$$P[\Theta \in [\tilde{\mu}, \tilde{\mu} + \pi)] \geq 1/2, \quad P[\Theta \in (\tilde{\mu} - \pi, \tilde{\mu}]] \geq 1/2. \tag{4.21}$$

A third circular measure of location is the *population modal direction*, $\breve{\mu}$. If Θ is discrete then $\breve{\mu}$ is that direction with the highest probability (in the polar representation of the distribution), whereas if Θ has a density, f, then $\breve{\mu}$ is that direction for which the polar representation of f is maximum. Given this definition, $\breve{\mu}$ need not be unique.

The fundamental circular measure of *concentration*, the mean resultant length, $\rho = \rho_1 \in [0, 1]$, defined in Equation (4.12), always exists. The mean resultant length is said to be a measure of concentration that is invariant under rotation as well as reflection, as its value is unchanged by rotating a distribution through an angle ϕ or reflecting it about any axis. When $\rho > 0$ (and therefore μ exists),

$$\bar{\beta}_1 = E[\sin(\Theta - \mu)] = E[\sin\Theta \cos\mu - \cos\Theta \sin\mu]$$
$$= E[\sin\Theta]\frac{\alpha_1}{\rho} - E[\cos\Theta]\frac{\beta_1}{\rho}$$
$$= \beta_1\frac{\alpha_1}{\rho} - \alpha_1\frac{\beta_1}{\rho} = 0. \tag{4.22}$$

A similar calculation leads to

$$\bar{\alpha}_1 = E[\cos(\Theta - \mu)] = \rho. \tag{4.23}$$

These two important identities are the population analogues of their sample counterparts in Equation (3.8).

The analogue of V, defined in Equation (3.13), is the *population circular variance*,

$$\upsilon = 1 - \rho \in [0, 1]. \tag{4.24}$$

Clearly, $\upsilon = 0$ when $\rho = 1$, i.e. when the distribution is a point distribution concentrated at $\Theta = \mu$. Similarly, when $\upsilon = 1$, $\rho = 0$. Whilst it is true that $\rho = 0$ for the continuous circular uniform distribution (see Section 4.3.3), ρ also equals 0 for any cyclically symmetric distribution (see Section 4.2.4), so $\upsilon = 0$ cannot be interpreted as necessarily implying that the whole of the population is highly scattered.

The population analogues of $\hat{\sigma}$ and $\hat{\delta}$, defined in Equations (3.14) and (3.16), are the *population circular standard deviation* and *dispersion*

$$\sigma = \{-2\log(1 - \upsilon)\}^{1/2} = \{-2\log\rho\}^{1/2} \in [0, \infty], \quad \delta = \frac{1 - \rho_2}{2\rho^2} \tag{4.25}$$

The basic population measures of circular skewness and kurtosis are the population analogues of $\bar{b}_2$ and $\bar{a}_2$ discussed in Section 3.5, namely the central sine and cosine moments $\bar{\beta}_2$ and $\bar{\alpha}_2$ defined in Equation (4.18). The population analogues of the standardized measures $\hat{s}$ and $\hat{k}$, defined in Equations (3.22) and (3.23), are the *population circular skewness* and *kurtosis*

$$s = \frac{\bar{\beta}_2}{(1 - \rho)^{3/2}}, \quad k = \frac{\bar{\alpha}_2 - \rho^4}{(1 - \rho)^2}. \tag{4.26}$$

4.2.4 *Symmetric Distributions*

The distribution of a linear random variable is said (unequivocally) to be 'symmetric' if there is a unique point, ξ say, on the real line about which the reflection of the distribution is identical to the original distribution. More precisely, then, such a distribution is *reflectively symmetric* (about ξ). As we shall see in Section 4.3, most of the classic models of circular statistics are also 'reflectively symmetric', but about a unique *axis*. The polar representation of a von Mises density in Fig. 4.1 illustrates this fact, the axis being that connecting the modal direction, π, and the origin, 0. More generally, if a circular distribution is reflectively symmetric about $\theta = \psi$ then it is also reflectively symmetric about $\theta = \psi + \pi$. Multimodal circular distributions can be reflectively symmetric about more than one axis, and the continuous circular uniform distribution (Section 4.3.3) is reflectively symmetric about any axis. This is an important point because the continuous circular uniform distribution is a limiting distribution of most circular models. If Θ has a distribution which is reflectively symmetric about ψ, and a density, $f(\theta)$, then

$$f(\theta - \psi) = f(\psi - \theta). \tag{4.27}$$

For unimodal reflectively symmetric distributions, the mean direction, μ, and the median direction, $\tilde{\mu}$, are identical, and equal to the modal direction, $\breve{\mu}$, when the latter is uniquely defined. More generally, for reflectively symmetric distributions with $\rho > 0$, all of the central sine moments, the $\bar{\beta}_p$, equal 0 and it follows that $\tau_{p,\mu} = \bar{\alpha}_p = \rho_p$. If, moreover, $\mu = 0$ then $\bar{\alpha}_p = \alpha_p$ and $\mu_p = 0$, and the Fourier expansion (4.16) simplifies to

$$f(\theta) = \frac{1}{2\pi}\left\{1 + 2\sum_{p=1}^{\infty} \alpha_p \cos p\theta\right\}. \tag{4.28}$$

There is another type of symmetry, however, that the distributions of circular random variables can exhibit but which, by definition, the distributions of linear random variables do not. A circular distribution is said to be *ℓ-fold symmetric* if its rotation through an angle $2\pi/\ell$ is identical to the original distribution. Thus, ℓ-fold symmetry can be interpreted as a form of *cyclic* symmetry. Obviously, all circular distributions are one-fold symmetric. *Antipodally symmetric* distributions, like the one with the density on the left of Fig. 4.2, are two-fold symmetric. Whilst that density is also reflectively symmetric (about two axes), an ℓ-fold symmetric distribution will generally not be. For example, the trimodal density portrayed on the right of Fig. 4.2 is that of a three-fold symmetric distribution which is not reflectively symmetric because the shape of the density around its modes is not symmetric. If a circular distribution is ℓ-fold symmetric for $\ell > 1$ then $\rho = 0$ and μ and any measures related to it are undefined. Moreover, the median and modal directions are non-unique.

4.2.5 *Large-sample Distribution of Key Circular Summaries*

In the initial phase of statistical inference, we will often be interested in performing inference for basic measures such as the mean direction, μ, the mean resultant length, ρ, and the second central sine and cosine moments, $\bar{\beta}_2$ and $\bar{\alpha}_2$. The obvious point estimates of these population measures are their sample analogues $\bar{\theta}$, $\bar{R}$, $\bar{b}_2$ and $\bar{a}_2$. However, in order to carry out other forms of inference, such as confidence interval construction and hypothesis testing, the sampling distribution of the point estimates must be derived or estimated. That sampling distribution generally depends on the distributional form of the population from which the data were sampled. However, here we consider the details of a very general large-sample result due to Pewsey (2004*a*).

Using the δ-method, Pewsey (2004*a*) derives the large-sample distribution of the random vector $\bar{\boldsymbol{\zeta}} = (\bar{\theta}, \bar{R}, \bar{b}_2, \bar{a}_2)^T$ for an underlying circular population with $\rho \in (0, 1)$. Note

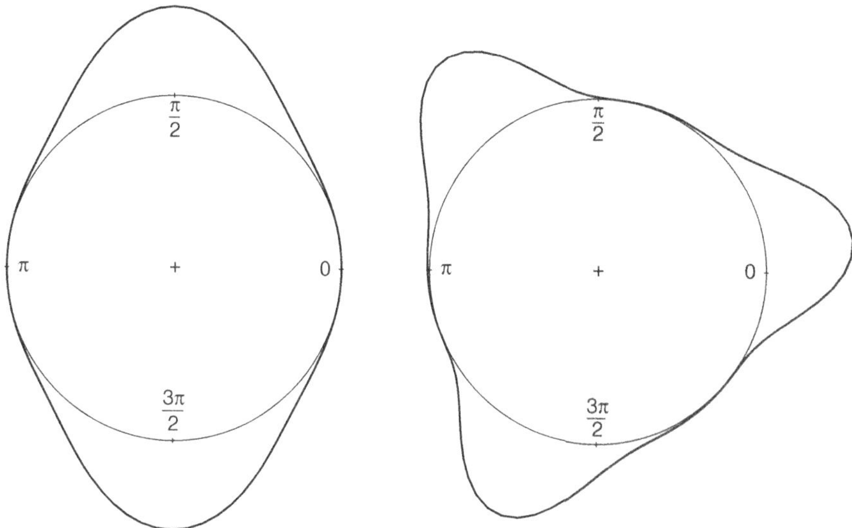

Figure 4.2 Polar representations of the densities of: an antipodally symmetric distribution (left); a three-fold symmetric distribution which is not reflectively symmetric (right)

that the result therefore does not apply to the types of distribution already discussed in this chapter, with $\rho = 0$, nor point distributions, for which $\rho = 1$. Pewsey shows that the large-sample distribution of $\bar{\boldsymbol{\zeta}}$ is asymptotically multivariate normal with mean vector $\boldsymbol{\xi}$ and variance covariance matrix $\boldsymbol{\Sigma}$, where, to $O(n^{-3/2})$,

$$\boldsymbol{\xi} = \left[\mu - \frac{\bar{\beta}_2}{2n\rho^2},\quad \rho + \frac{(1-\bar{\alpha}_2)}{4n\rho},\quad \bar{\beta}_2 + \frac{1}{n\rho}\left(-\bar{\beta}_3 - \frac{\bar{\beta}_2}{\rho} + \frac{2\bar{\alpha}_2\bar{\beta}_2}{\rho^3}\right),\right.$$
$$\left.\bar{\alpha}_2 + \frac{1}{n}\left\{1 - \frac{\bar{\alpha}_3}{\rho} - \frac{\bar{\alpha}_2(1-\bar{\alpha}_2) + \bar{\beta}_2^2}{\rho^2}\right\}\right]^T, \tag{4.29}$$

and $2n\rho^2\boldsymbol{\Sigma} = (\upsilon_{ij} : i, j = 1, \ldots, 4)$, where

$$\begin{aligned}
&\upsilon_{11} = 1 - \bar{\alpha}_2,\quad \upsilon_{12} = \upsilon_{21} = \rho\bar{\beta}_2,\\
&\upsilon_{13} = \upsilon_{31} = \rho^2 - \rho\bar{\alpha}_3 - 2\bar{\alpha}_2(1-\bar{\alpha}_2),\\
&\upsilon_{14} = \upsilon_{41} = \rho\bar{\beta}_3 + 2\bar{\beta}_2(1-\bar{\alpha}_2),\\
&\upsilon_{22} = \rho^2(1 - 2\rho^2 + \bar{\alpha}_2),\quad \upsilon_{23} = \upsilon_{32} = -2\rho^3\bar{\beta}_2 + \rho^2\bar{\beta}_3 - 2\rho\bar{\alpha}_2\bar{\beta}_2,\\
&\upsilon_{24} = \upsilon_{42} = \rho^3(1 - 2\bar{\alpha}_2) + \rho^2\bar{\alpha}_3 + 2\rho\bar{\beta}_2^2,\\
&\upsilon_{33} = \rho^2(1 - 4\bar{\alpha}_2 - 2\bar{\beta}_2^2 - \bar{\alpha}_4) + 4\rho\bar{\alpha}_2\bar{\alpha}_3 + 4\bar{\alpha}_2^2(1-\bar{\alpha}_2),\\
&\upsilon_{34} = \rho^2\left\{2(1-\bar{\alpha}_2)\bar{\beta}_2 + \bar{\beta}_4\right\} - 2\rho(\bar{\alpha}_2\bar{\beta}_3 + \bar{\beta}_2\bar{\alpha}_3) - 4\bar{\alpha}_2\bar{\beta}_2(1-\bar{\alpha}_2),\\
&\upsilon_{43} = \upsilon_{34},\quad \upsilon_{44} = \rho^2(1 - 2\bar{\alpha}_2^2 + \bar{\alpha}_4) + 4\rho\bar{\beta}_2\bar{\beta}_3 + 4\bar{\beta}_2^2(1-\bar{\alpha}_2).
\end{aligned} \tag{4.30}$$

To $O(n^{-3/2})$, then, $\bar{\theta}$, $\bar{R}$, $\bar{b}_2$ and $\bar{a}_2$ are biased estimates of μ, ρ, $\bar{\beta}_2$ and $\bar{\alpha}_2$, their biases, as well as their variances and covariances, depending on the sample size, n, the mean resultant length, ρ, and the second, third and fourth central sine and cosine moments of the population.

In Section 5.3 we use a plug-in estimate of the asymptotic distribution to construct confidence intervals and carry out hypothesis tests for the population measures μ, ρ, $\bar{\beta}_2$ and $\bar{\alpha}_2$.

For the interested reader, a more extensive treatment of circular distribution theory is provided by Mardia and Jupp (1999, Chapter 4).

4.3 Circular Models

In this section we consider most of the classical models for circular data as well as various more flexible models that have been proposed more recently. We start with a description of various general methods that can be used to generate circular distributions, before describing the main properties of the discrete and continuous circular uniform distributions. We progress to reflectively symmetric unimodal models, more flexible families of distributions capable of modelling features such as asymmetry, varying levels of kurtosis, and multimodality, and finally models for toroidal and cylindrical data.

Throughout, we provide **R** commands and functions for calculating the values of probability density, distribution and quantile functions as well as for simulating random variates. The latter capability is an essential component of computer-intensive methods of inference such as the parametric bootstrap.

4.3.1 *General Approaches for Generating Circular Distributions*

There are various general methods that can be used to obtain circular distributions. Perhaps the simplest is *perturbation*. In this approach, the density of an existing circular density is multiplied by some function chosen to ensure that their product is also a *bone fide* circular density. The cardioid and sine-skewed distributions of Sections 4.3.4 and 4.3.11 are examples of this type of construction.

A second approach is *wrapping*. Consider a random variable, X, defined on the real line. Wrapping the distribution of X around the circumference of the unit circle produces a circular random variable

$$\Theta = X \; (\mathrm{mod}\; 2\pi). \tag{4.31}$$

An important practical implication of this relation between X and Θ is that if it is possible to simulate variates from the distribution of X then the simulation of variates from the wrapped distribution of Θ is trivial. Denoting the distribution function of X by $F_X(x)$, the circular distribution function of Θ is given by

$$F_\Theta(\theta) = \sum_{k=-\infty}^{\infty} \{F_X(\theta + 2\pi k) - F_X(2\pi k)\}, \quad 0 \leq \theta \leq 2\pi. \tag{4.32}$$

If, moreover, X has a probability density function, $f_X(x)$, then Θ has probability density function,

$$f_\Theta(\theta) = \sum_{k=-\infty}^{\infty} f_X(\theta + 2\pi k). \tag{4.33}$$

It transpires that this last result is a particularly unappealing property because, for all known continuous circular distributions apart for the wrapped Cauchy distribution (see Section 4.3.6), the infinite sum in (4.33) does not simplify to a closed form. Consequently, the densities of wrapped distributions are generally cumbersome to deal with.

Three other important properties of wrapped distributions are:

(i) If X has characteristic function $\psi_X(t) = E[e^{itX}]$ then the trigonometric moments in the characteristic function, $\{\tau_{p,0} : p = 0, \pm 1, \pm 2, \ldots\}$, of $\Theta = X \pmod{2\pi}$ are given by

$$\tau_{p,0} = \psi_X(p). \tag{4.34}$$

Thus, if we know the characteristic function of X then obtaining the trigonometric moments of Θ is trivial.

(ii) If $\psi_X(t)$ is integrable then X has a density, $f_X(x)$, and $\Theta = X \pmod{2\pi}$ has a density which can be represented as

$$f_\Theta(\theta) = \frac{1}{2\pi}\left\{1 + 2\sum_{p=1}^{\infty}(\alpha_p \cos p\theta + \beta_p \sin p\theta)\right\}, \tag{4.35}$$

where $\psi_X(p) = \alpha_p + i\beta_p$. Combined with property (i), this result implies that it will be trivial to write down a series expansion of the density of Θ if we know what the characteristic function of X is. The downside is that the infinite series in (4.35) will not, in general, simplify to a closed form.

(iii) If X and Y are two random variables defined on the line and $\Theta = X \pmod{2\pi}$ and $\Psi = Y \pmod{2\pi}$ are their wrapped counterparts then

$$\Phi = (X + Y) \pmod{2\pi} = \Theta + \Psi \pmod{2\pi}. \tag{4.36}$$

In words, the wrapped counterpart of their sum equals the sum of their wrapped counterparts (mod 2π).

In Sections 4.3.6 and 4.3.7 we describe the main properties of the two most basic wrapped distributions; the wrapped Cauchy and wrapped normal distributions. Later, in

Section 4.3.15, we mention other wrapped distributions including the wrapped stable family for which the wrapped Cauchy and wrapped normal distributions are special cases. A more formal mathematical treatment of wrapped distributions is given by Mardia and Jupp (1999, Section 3.5.7).

A third general method of construction is to apply *transformation of argument*, or *scale*, to some existing density, $f(\theta)$, replacing its argument θ by some function of it. Various families of distributions generated in this way will be considered in Sections 4.3.10, 4.3.12 and 4.3.13.

A further type of transformation that can be applied to a distribution defined on the unit circle is the so-called *Möbius transformation*. The model of Kato and Jones (2010), referred to briefly in Section 4.3.15, is obtained by applying this form of transformation to the von Mises distribution.

Finally, *projected* or *offset* distributions can be derived by first obtaining the polar representation, $f(r,\theta)$, of a bivariate linear density, $f(x,y)$, and then integrating over r to obtain the marginal density of $\Theta, f(\theta)$. For details of the application of this approach to the bivariate normal distribution, the interested reader is referred to Mardia (1972, Section 3.4.7), Mardia and Jupp (1999, Section 3.5.6) and Jammalamadaka and SenGupta (2001, Section 2.2.5).

4.3.2 *Discrete Circular Uniform Distribution*

The random variable Θ follows a *discrete circular uniform distribution* on m points if it has a probability distribution of the form

$$P\left(\Theta = \xi + \frac{2\pi q}{m}\right) = \frac{1}{m}, \quad q = 0, 1, \ldots, m-1. \tag{4.37}$$

When $m = 1$, (4.37) defines a *point distribution* with all the probability mass located at $\theta = \xi$. For such a distribution, $\mu = \tilde{\mu} = \breve{\mu} = \xi$. Moreover, $\alpha_p = \cos p\xi$ and $\beta_p = \sin p\xi$, and hence $\tau_{p,0} = \cos p\xi + i \sin p\xi$, $\mu_p = p\xi \pmod{2\pi}$, $\rho_p = 1, \bar{\alpha}_p = 1, \bar{\beta}_p = 0$ and $\tau_{p,\mu} = 1$.

An antipodally symmetric distribution, with half the probability mass located at $\theta = \xi$ and the other half at $\theta = \xi + \pi$, is obtained when $m = 2$. More generally, when $m > 2$, Equation (4.37) defines a distribution with identical probabilities of $1/m$ located at m equally spaced points on the unit circle. When $m > 1$, the mean direction is undefined and the median and modal directions are non-unique. For $p = 0 \pmod{m}$, the trigonometric moments are the same as those given above for the case $m = 1$. Otherwise, $\alpha_p = \beta_p = \tau_{p,0} = \rho_p = 0$, μ_p is undefined and hence so are all those measures based upon it.

The function **dcuSim** below can be used to simulate a sample of size n from a discrete uniform distribution with supplied values of ξ and m.

```
dcuSim <- function(n, xi, m) {
dcuniloc <- seq(0:m-1) ; dcuniloc <- dcuniloc*2*pi/m + xi
dcusamp <- sample(dcuniloc, n, replace=TRUE)
dcusamp <- circular(dcusamp) ; return(dcusamp)
}
```

Using **dcuSim** within the commands:

```
library(circular) ; n <- 100 ; xi <- pi/2 ; m <- 10
dcusamp <- dcuSim(n, xi, m)
```

simulates a random sample of size $n = 100$ from the discrete circular uniform distribution with $\xi = \pi/2$ and $m = 10$ and assigns it to the object **dcusamp**.

4.3.3 *Continuous Circular Uniform Distribution*

The *continuous circular uniform distribution*, with density

$$f(\theta) = \frac{1}{2\pi}, \tag{4.38}$$

is the most fundamental of circular models, appropriate when no direction is any more likely than any other. It is the unique circular distribution that is invariant under both rotation and reflection. Clearly, and importantly, the density (4.38) involves no parameters. As all directions are equally likely, this distribution characterizes *isotropy* or *circular randomness.*

Integrating (4.38), the distribution function of Θ is simply

$$F(\theta) = \frac{\theta}{2\pi}. \tag{4.39}$$

For $\phi \leq \psi \leq \phi + 2\pi$, then,

$$P(\phi < \Theta \leq \psi) = F(\psi) - F(\phi) = \frac{\psi - \phi}{2\pi}.$$

Inverting the distribution function, the quantile function is

$$Q(u) = F^{-1}(u) = 2\pi u, \quad 0 \leq u \leq 1. \tag{4.40}$$

The trigonometric moments about the zero direction are given by

$$\tau_{p,0} = \begin{cases} 1, & p = 0, \\ 0, & p \neq 0. \end{cases} \tag{4.41}$$

Thus, $\rho = 0$ and hence μ, and all those measures based upon it, are undefined. Because the polar representation of (4.38) is ring-like in shape, $\tilde{\mu}$ and $\check{\mu}$ are clearly non-unique.

As Mardia and Jupp (1999, Section 4.3.1) show, the distribution of the sum of n independent and identically distributed random variables from a circular distribution tends to a circular uniform distribution as $n \to \infty$.

Employing **R**'s **circular** package, a polar representation of the density of the continuous circular uniform distribution can be obtained using the command:

```
curve.circular(dcircularuniform, join=TRUE, ylim=c(-1.05, 1.05), lwd=2)
```

and a random sample of size $n = 100$ simulated from it using the command:

```
ccusamp <- rcircularuniform(100, control.circular=list(units="radians"))
```

The function **curve.circular** draws a curve as defined by the expression in its first argument. In this case, **dcircularuniform** is a function within the **circular** package for calculating values of the continuous circular uniform density. The setting **join=TRUE** indicates that the first and last points of the function should be joined; **ylim** and **lwd** control the scale of the vertical axis and the line width, respectively. **rcircularuniform** is a function within the **circular** package that generates random variates from the continuous circular uniform distribution. Its first argument is the number of observations to be generated, and the second argument specifies the attributes of the resulting object.

Values of the distribution function, $F(\theta)$, can be computed using the function:

```
ccuDF <- function(theta) { theta/(2*pi) }
```

For instance, $F(2) = 2/(2\pi) = 0.3183$, can be calculated using the command:

```
ccudf2 <- ccuDF(2)
```

Inverting the distribution function, the function **ccuQF** computes the value of the quantile function, $Q(u)$, of the continuous circular uniform distribution for a supplied value of u.

```
ccuQF <- function(u) { u*(2*pi) }
```

For example, $Q(2/(2\pi)) = 2$ is returned using the command:

```
ccuQF(ccudf2)
```

As for all of the new functions introduced in this book, here the object **theta** is assumed to be a *linear* (not a *circular*) one containing values in $[0, 2\pi)$, and **u** to be a (linear) one containing values in $[0, 1]$.

Circular uniformity (or isotropy, or circular randomness) is the most basic dividing hypothesis in circular statistics and, as we shall see in Section 5.1, numerous tests have been developed to test for it. If circular uniformity cannot be rejected there is no need to search any further for a more complicated model to describe our data. If it is rejected then the parametric models described below provide potential alternatives.

4.3.4 *Cardioid Distribution*

The *cardioid*, or *cosine*, distribution, introduced by Jeffreys (1948, p. 302), has a density which can be considered to arise from *cosine perturbation* of the continuous circular uniform density. To ensure identifiability, we specify its density as

$$f(\theta) = \frac{1}{2\pi}\{1 + 2\rho\cos(\theta - \mu)\}, \quad \rho \in [0, 1/2], \tag{4.42}$$

the parameter ρ representing the distribution's mean resultant length. The uniform distribution is obtained when $\rho = 0$. Otherwise, when $0 < \rho \leq 1/2$, the distribution is unimodal and reflectively symmetric about $\mu = \tilde{\mu} = \check{\mu}$. For $0 < \rho < 0.1$, the distribution corresponds to a mild departure from uniformity with slightly more density in the neighbourhood of μ than around the antimode $\mu + \pi$.

Integrating (4.42),

$$F(\theta) = \frac{1}{2\pi}\{\theta + 2\rho[\sin\mu + \sin(\theta - \mu)]\}, \tag{4.43}$$

and hence for $\phi \leq \psi \leq \phi + 2\pi$,

$$P(\phi < \Theta \leq \psi) = F(\psi) - F(\phi) = \frac{1}{2\pi}\{(\psi - \phi) + 2\rho[\sin(\psi - \mu) - \sin(\phi - \mu)]\}.$$

There is no closed-form expression for the quantile function $Q(u)$ but its values can be obtained by inverting the distribution function numerically.

The trigonometric moments about the zero direction are given by

$$\tau_{p,0} = \begin{cases} \rho e^{i\mu}, & p = 1, \\ 0, & p \neq 1. \end{cases} \tag{4.44}$$

In what follows we consider computation using **R**'s **circular** package for a cardioid distribution with $\mu = \pi/2$ and $\rho = 0.3$. A polar representation of its density can be obtained using the commands:

```
mu <- circular(pi/2) ; rho <- 0.3
curve.circular(dcardioid(x, mu, rho), join=TRUE, ylim=c(-1.2,1.2), lwd=2)
```

and a random sample of size $n = 100$ simulated from it, using a simple acceptance–rejection method, employing the command:

```
cardsamp <- rcardioid(100, mu, rho, control.circular=list(units="radians"))
```

The functions **cardioidDF** and **cardioidQF** below can be used to compute values of the distribution function, $F(\theta)$, and the quantile function, $Q(u)$, respectively, for a cardioid distribution with specified values of μ and ρ. Numerical inversion of $F(x)$, using **R**'s root-finding function **unitroot**, is used when computing $Q(u)$.

```
cardioidDF <- function(theta, mu, rho) {
dfval <- (theta+2*rho*(sin(mu)+sin(theta-mu)))/(2*pi) ; return(dfval)
}

cardioidQF <- function(u, mu, rho) {
eps <- 10*.Machine$double.eps
if (u <= eps) {theta <- 0 ; return(theta)} else
if (u >= 1-eps) {theta <- 2*pi-eps ; return(theta)}
else {
roottol <- .Machine$double.eps**(0.6)
qzero <- function(x) { y <- cardioidDF(x, mu, rho)-u ; return(y) }
res <- uniroot(qzero, lower=0, upper=2*pi-eps, tol=roottol)
theta <- res$root ; return(theta) }
}
```

For example, the values of $F(\pi) = 0.6909859$ and $Q(0.6909859) = \pi$ can be computed using the commands:

```
carddfpi <- cardioidDF(pi, pi/2, rho) ; cardioidQF(carddfpi, pi/2, rho)
```

A construction leading to the cardioid distribution, involving marbles falling onto a tilted tray, is described by Fisher (1993, Section 3.3.2).

4.3.5 *Cartwright's Power-of-Cosine Distribution*

A model related to the cardioid distribution is the *power-of-cosine* distribution due to Cartwright (1963). Its density is given by

$$f(\theta) = \frac{2^{-1+1/\zeta}\,\Gamma^2(1+1/\zeta)}{\pi\,\Gamma(1+2/\zeta)}(1+\cos(\theta-\mu))^{1/\zeta}, \tag{4.45}$$

where Γ denotes the gamma function and $\zeta > 0$ is a parameter controlling the concentration and, as a consequence, the shape of the distribution.

The cardioid distribution with $\rho = 1/2$ is obtained when $\zeta = 1$. As $\zeta \to 0$ the density becomes increasingly more concentrated about the mean direction, μ, whilst as $\zeta \to \infty$ the density tends to that of the continuous circular uniform distribution (apart from at the antimode $\mu + \pi$). These features are illustrated by the three densities with mean direction $\mu = \pi/2$ and $\zeta = 0.1, 1, 10$ portrayed in Fig. 4.3. The plot was generated using **R**'s **circular** package and the following commands:

```
mu <- circular(pi/2) ; zeta <- 1 ; theta <- circular(seq(0, 2*pi, by=pi/3600))
curve.circular(dcarthwrite(x, mu, zeta), join=TRUE, ylim=c(-1, 1.8), cex=0.7, lwd=2)
zeta <- 10 ; y <- dcarthwrite(theta, mu, zeta) ; lines(theta, y, lty=2, lwd=2)
zeta <- 0.1 ; y <- dcarthwrite(theta, mu, zeta) ; lines(theta, y, lty=4, lwd=2)
```

Note the misspelling of Cartwright's name in the **dcarthwrite** function and throughout the documentation for the **circular** package.

In what follows we consider computation in **R** for a power-of-cosine distribution with $\mu = \pi/2$ and $\zeta = 0.1$. There are no in-built functions for computing values of the distribution function or the quantile function of a power-of-cosine distribution in the **circular** package. Below we first define the function **CartwrightPDF** for computing values of the density (4.45) and then use it, together with numerical integration, within the function **CartwrightDF** written to calculate values of $F(\theta)$. In turn, **CartwrightDF** is called within the function **CartwrightQF** programmed to calculate values of $Q(u)$ via numerical inversion of $F(\theta)$.

```
CartwrightPDF <- function(theta, mu, zeta) {
pdfval <- (2**(1/zeta-1))*((gamma(1+1/zeta))**2)
pdfval <- pdfval*((1+cos(theta-mu))**(1/zeta))/(pi*gamma(1+2/zeta))
return(pdfval)
}

CartwrightDF <- function(theta, mu, zeta) {
eps <- 10*.Machine$double.eps
if (theta <= eps) { dfval <- 0 ; return(dfval) } else
if (theta >= 2*pi-eps) { dfval <- 1 ; return(dfval) }
```

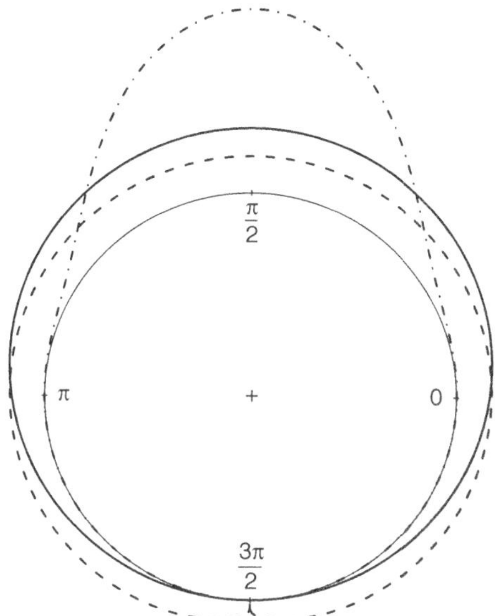

Figure 4.3 Polar representations of power-of-cosine densities with $\mu = \pi/2$ and: $\zeta = 1$ (solid); $\zeta = 10$ (dashed); $\zeta = 0.1$ (dot-dashed)

```
else {
dfval <- integrate(CartwrightPDF, mu=mu, zeta=zeta, lower=0, upper=theta)$value
return(dfval) }
}

CartwrightQF <- function(u, mu, zeta) {
eps <- 10*.Machine$double.eps
if (u <= eps) {theta <- 0 ; return(theta)} else
if (u >= 1-eps) {theta <- 2*pi-eps ; return(theta)}
else {
roottol <- .Machine$double.eps**(0.6)
qzero <- function(x) { y <- CartwrightDF(x, mu, zeta)-u ; return(y) }
res <- uniroot(qzero, lower=0, upper=2*pi-eps, tol=roottol)
theta <- res$root ; return(theta) }
}
```

Having defined these three new functions it is now simple to compute, for instance, $F(3\pi/4) = 0.9642$ and $Q(0.9642) = 3\pi/4$ using the following commands:

```
theta <- 3*pi/4
cwdfval <- CartwrightDF(theta, mu, zeta) ; CartwrightQF(cwdfval, mu, zeta)
```

Nor is there a function in the **circular** package for simulating random variates from Cartwright's power-of-cosine distribution. The function **CartwrightSim** below applies an acceptance–rejection algorithm with a rectangular envelope to simulate a random sample of size n from a power-of-cosine distribution with specified values of μ and ζ. For all but the most concentrated of cases, it is more efficient than an alternative function available from the website which makes use of the **CartwrightQF** function to implement inverse transform sampling.

```
CartwrightSim <- function(n, mu, zeta) {
fmax <- CartwrightPDF(mu, mu, zeta) ; theta <- 0
for (j in 1:n) { stopgo <- 0
while (stopgo == 0) {
u1 <- runif(1, 0, 2*pi) ; pdfu1 <- CartwrightPDF(u1, mu, zeta)
u2 <- runif(1, 0, fmax)
if (u2 <= pdfu1) { theta[j] <- u1 ; stopgo <- 1 }
}}
return(theta)
}
```

Note that **CartwrightSim** returns a linear data object (not a circular one), containing values in $[0, 2\pi)$. This is also true for all subsequent functions with names ending in **Sim**. Any such object can be converted to a circular one using **R**'s **circular** function. Using this new function, we can simulate, for example, a sample of size $n = 100$ from the power-of-cosine distribution under consideration using the commands:

```
n <- 100 ; cartsamp <- CartwrightSim(n, mu, zeta)
```

The power-of-cosine distribution was first proposed by Cartwright (1963) as a model for the directional spectra of ocean waves.

4.3.6 *Wrapped Cauchy Distribution*

A Cauchy random variable defined on the real line, X, has density

$$f(x) = \frac{1}{\pi} \frac{\gamma}{\gamma^2 + (x - \xi)^2},$$

where $-\infty < \xi < \infty$ is the median and mode but, famously, not the mean, of the distribution, and $\gamma > 0$ is a scale parameter. Its characteristic function is $\psi_X(t) = e^{it\xi - \gamma|t|}$, and hence $\psi_X(p) = \alpha_p + i\beta_p$ with $\alpha_p = e^{-\gamma|p|}\cos(p\xi)$ and $\beta_p = e^{-\gamma|p|}\sin(p\xi)$. It follows from (4.35) that the *wrapped Cauchy distribution*, of the random variable $\Theta = X \pmod{2\pi}$, has density

$$f(\theta) = \frac{1}{2\pi}\left\{1 + 2\sum_{p=1}^{\infty} \rho^p \cos p(\theta - \mu)\right\}, \tag{4.46}$$

where $\rho = e^{-\gamma}$ is the mean resultant length and $\mu = \xi \pmod{2\pi}$ is the mean direction. Given (4.34), the trigonometric moments of Θ can be expressed as

$$\tau_{p,0} = \rho^{|p|}\cos(p\mu) + i\rho^{|p|}\sin(p\mu) = \rho^{|p|}e^{ip\mu}.$$

The infinite sum in (4.46) is the real part of the geometric series $\sum_{p=1}^{\infty} \rho^p e^{-ip(\theta-\mu)}$, which is easily shown to equal

$$\frac{\rho\{\cos(\theta - \mu) - \rho\}}{1 + \rho^2 - 2\rho\cos(\theta - \mu)}.$$

Hence (4.46) simplifies to

$$f(\theta) = \frac{1}{2\pi}\frac{1 - \rho^2}{1 + \rho^2 - 2\rho\cos(\theta - \mu)}. \tag{4.47}$$

The continuous circular uniform distribution is obtained when $\rho = 0$. More generally, for $\rho > 0$ the polar representation of the density is unimodal and reflectively symmetric about μ. As the densities on the left of Fig. 4.4 illustrate, the density around the antimode is substantial even for $\rho = 0.75$.

In what follows we consider computation in **R** for a wrapped Cauchy distribution with $\mu = \pi/2$ and $\rho = 0.75$. Within the **circular** package, a polar representation of its density can be produced using the commands:

```
mu <- circular(pi/2) ; rho <- 0.75
curve.circular(dwrappedcauchy(x, mu, rho), join=TRUE, xlim=c(-1, 2), lwd=2)
```

and a random sample of size $n = 100$ simulated from it using the command:

```
wcauchysamp <- rwrappedcauchy(100, mu, rho, control.circular=list(units="radians"))
```

The algorithm implemented in the function **rwrappedcauchy** simulates a (linear) Cauchy random variate and then computes its value mod 2π.

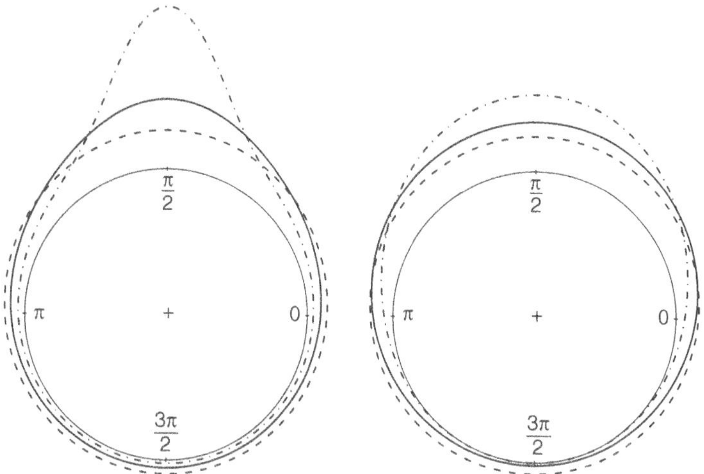

Figure 4.4 Polar representations of wrapped Cauchy (left) and wrapped normal (right) densities with $\mu = \pi/2$ and $\rho = 0.25$ (dashed), $\rho = 0.5$ (solid), $\rho = 0.75$ (dot-dashed)

There are no functions for calculating values of the distribution and quantile functions available in the **circular** package. Following the same approach as presented in Section 4.3.5, below we provide functions to compute values of the density, distribution and quantile functions, respectively:

```
wCauchyPDF <- function(theta, mu, rho) {
pdfval <- (1-rho**2)/((1+rho**2-2*rho*cos(theta-mu))*(2*pi))
return(pdfval)
}
wCauchyDF <- function(theta, mu, rho) {
eps <- 10*.Machine$double.eps
if (theta <= eps) { dfval <- 0 ; return(dfval) } else
if (theta >= 2*pi-eps) { dfval <- 1 ; return(dfval) }
else {
dfval <- integrate(wCauchyPDF, mu=mu, rho=rho, lower=0, upper=theta)$value
return(dfval) }
}
wCauchyQF <- function(u, mu, rho) {
eps <- 10*.Machine$double.eps
if (u <= eps) {theta <- 0 ; return(theta)} else
if (u >= 1-eps) {theta <- 2*pi-eps ; return(theta)}
else {
roottol <- .Machine$double.eps**(0.6)
qzero <- function(x) { y <- wCauchyDF(x, mu, rho)-u ; return(y) }
res <- uniroot(qzero, lower=0, upper=2*pi-eps, tol=roottol)
theta <- res$root ; return(theta) }
}
```

With the aid of these new functions, the values of $F(\pi/6) = 0.0320$ and $Q(0.0320) = \pi/6$, for example, can be calculated using the commands:

```
theta <- pi/6 ; mu <- pi/2
wcdfval <- wCauchyDF(theta, mu, rho) ; wCauchyQF(wcdfval, mu, rho)
```

McCullagh (1996) shows that a wrapped Cauchy random variable can be generated by Möbius transformation of a continuous circular uniform random variable. The wrapped Cauchy distribution is a special case of the Jones–Pewsey, wrapped t and wrapped stable families considered in Sections 4.3.9 and 4.3.15.

4.3.7 *Wrapped Normal Distribution*

Wrapping a normal random variable with mean $\xi \in (-\infty, \infty)$ and variance $\sigma^2 > 0$, $X \sim N(\xi, \sigma^2)$, around the circumference of the unit circle produces a circular random variable, Θ, with a *wrapped normal distribution*, the density of which is

$$f(\theta) = \frac{1}{(-4\pi \log \rho)^{1/2}} \sum_{k=-\infty}^{\infty} \exp\left\{\frac{(\theta - \mu + 2\pi k)^2}{4 \log \rho}\right\}, \tag{4.48}$$

where $\rho = e^{-\sigma^2/2} \in (0, 1)$ is the mean resultant length and $\mu = \xi \pmod{2\pi}$ is the mean direction. As the characteristic function of X is $\psi_X(t) = \exp(it\xi - t^2\sigma^2/2)$, using (4.34) we obtain,

$$\tau_{p,0} = \rho^{p^2} e^{ip\mu}, \quad \alpha_p = \rho^{p^2} \cos(p\mu), \quad \beta_p = \rho^{p^2} \sin(p\mu). \tag{4.49}$$

Applying (4.35), an alternative representation of the density of Θ is therefore given by

$$f(\theta) = \frac{1}{2\pi} \left\{ 1 + 2 \sum_{p=1}^{\infty} \rho^{p^2} \cos p(\theta - \mu) \right\}. \tag{4.50}$$

In practice, the wrapped normal density is approximated reasonably well by the first three terms of (4.50) when $\sigma^2 \geq 2\pi$, or by the term for $k = 0$ of (4.48) when $\sigma^2 \leq 2\pi$.

The polar representation of a wrapped normal distribution is unimodal and reflectively symmetric about μ, tending to the continuous circular uniform distribution as $\rho \to 0$ and a point distribution concentrated at μ as $\rho \to 1$. As a comparison of the densities on the left and right of Fig. 4.4 verifies, for a given mean resultant length, wrapped normal densities are generally less concentrated about the mean direction, and have less density around the antimode, than their wrapped Cauchy counterparts.

Values of the distribution function for $0 \leq \theta < 2\pi$ can be computed using numerical integration of either (4.48) or (4.50), and values of the quantile function for $0 \leq u \leq 1$ via numerical inversion of the distribution function.

To illustrate the functionality of **R**'s **circular** package, in what follows we consider computation for a wrapped normal density with $\mu = \pi/2$ and $\rho = 0.75$. A polar representation of its density can be obtained using the commands:

```
mu <- circular(pi/2) ; rho <- 0.75
curve.circular(dwrappednormal(x, mu, rho), join=TRUE, xlim=c(-1, 2), lwd=2)
```

and a random sample of size $n = 100$ simulated from it using the command:

```
wnormsamp <- rwrappednormal(100, mu, rho, control.circular=list(units="radians"))
```

The algorithm used in the function **rwrappednormal** simulates a linear normal random variate and then computes its value mod 2π. The value of distribution function for $\theta = \pi/6$, $F(\pi/6) = 0.0645$, can be computed using the command:

```
pwrappednormal(circular(pi/6), mu, rho, from=circular(0))
```

and the value of the quantile function for $u = 0.5$, $Q(0.5) = 1.6073$, using the command:

```
qwrappednormal(0.5, mu, rho, from=circular(0), tol=.Machine$double.eps**(0.6))
```

Mardia and Jupp (1999, page 51) describe two constructions involving Brownian motion on the circle which lead to the wrapped normal distribution.

4.3.8 Von Mises Distribution

The *von Mises distribution* has density

$$f(\theta) = \frac{1}{2\pi I_0(\kappa)} e^{\kappa \cos(\theta-\mu)}, \tag{4.51}$$

where $\kappa \geq 0$ is the so-called *concentration parameter,*

$$I_p(\kappa) = \frac{1}{2\pi} \int_0^{2\pi} \cos p\theta \, e^{\kappa \cos\theta} d\theta \tag{4.52}$$

is the modified Bessel function of the first kind and order p, and, when $\kappa > 0$, μ denotes the mean direction. The apparent simplicity of (4.51), especially when compared with a wrapped density like (4.48), is rather misleading. Rather than involving an infinite sum, (4.51) requires the use of numerical integration to evaluate $I_0(\kappa)$. When required, values of (4.52) can be computed using the **I.p** function available in **R**'s **circular** package.

The continuous circular uniform distribution is obtained when $\kappa = 0$, and the distribution tends to a point distribution centred on μ as $\kappa \to \infty$. When $\kappa > 0$, a polar representation of the density is unimodal and reflectively symmetric about $\mu = \tilde{\mu} = \check{\mu}$. Because of this symmetry, for $\kappa > 0$, $\bar{\beta}_p = 0$ and

$$\bar{\alpha}_p = \frac{1}{2\pi I_0(\kappa)} \int_0^{2\pi} \cos p(\theta - \mu) e^{\kappa \cos(\theta-\mu)} d\theta = \frac{I_p(\kappa)}{I_0(\kappa)} = A_p(\kappa). \tag{4.53}$$

So, when $\kappa > 0$, $\tau_{p,\mu} = A_p(\kappa)$ and it follows, using (4.19), that

$$\tau_{p,0} = \tau_{p,\mu} e^{ip\mu} = A_p(\kappa) e^{ip\mu}.$$

Thus, when $\kappa > 0$,

$$\alpha_p = A_p(\kappa) \cos p\mu, \quad \beta_p = A_p(\kappa) \sin p\mu, \quad \rho_p = A_p(\kappa), \quad \mu_p = p\mu.$$

In particular, the mean resultant length is $\rho = \bar{\alpha}_1 = A_1(\kappa)$, and it can be shown that $\bar{\alpha}_2 = A_2 = 1 - 2A_1(\kappa)/\kappa$. Values of A_1 can be computed using the **A1** function available within **R**'s **circular** package.

Considering the polar representations of the densities portrayed in Fig. 4.5, it is evident that the cardioid, wrapped Cauchy, wrapped normal and von Mises densities with the same mean direction and mean resultant length are all very similar when the mean resultant length, ρ, is small. As ρ increases, the disparity between the wrapped Cauchy density and the other three densities increases. Of the three, the wrapped normal density provides the closest approximation to the von Mises density, although it generally has marginally less density around the mode and slightly more density around the shoulders than the von Mises. Clearly, for large values of ρ, the wrapped Cauchy provides a very poor approximation to the von Mises. In fact, Pewsey *et al.* (2007) show that for $\rho = 0.05$ the best wrapped t (see Section 4.3.15) approximation to the von Mises corresponds to one with

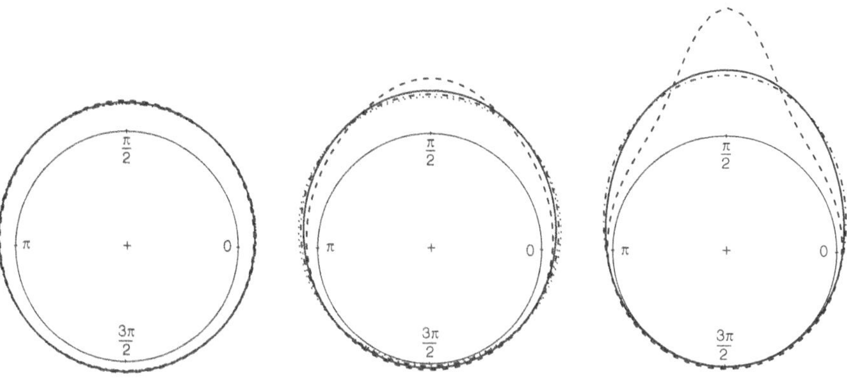

Figure 4.5 Polar representations of the densities of von Mises (solid), cardioid (dotted), wrapped normal (dot-dashed) and wrapped Cauchy (dashed) distributions with $\mu = \pi/2$ and mean resultant lengths of $\rho = 0.25$ (left), $\rho = 0.5$ (centre) and $\rho = 0.75$ (right). A cardioid density does not appear in the plot on the right as there $\rho > 0.5$

$\nu = 1.99$ degrees of freedom; whilst for $\rho = 0.75$, $\nu = 8.53$. These results square with the above observations, the wrapped Cauchy distribution being a wrapped t distribution with one degree of freedom. They also indicate that the wrapped normal approximation to the von Mises distribution, corresponding to a wrapped t distribution with $\nu = \infty$ degrees of freedom, can generally be improved upon.

For $0 \leq \theta < 2\pi$, values of the distribution function of the von Mises distribution are given by

$$F(\theta) = \frac{1}{2\pi I_0(\kappa)} \int_0^{\theta} e^{\kappa \cos(\phi - \mu)} \, d\phi, \tag{4.54}$$

the calculation of both the Bessel function and the integral requiring quadrature.

In what follows we illustrate computation using the functionality of **R**'s **circular** package for a von Mises density with $\mu = \pi/2$ and $\kappa = 2.37$ (and hence $\rho = 0.75$). A polar representation of its density can be obtained using the commands:

```
mu <- circular(pi/2) ; kappa <- 2.37
curve.circular(dvonmises(x, mu, kappa), join=TRUE, xlim=c(-1, 2), lwd=2)
```

and a random sample of size $n = 100$ simulated from it using the command:

```
vmsamp <- rvonmises(100, mu, kappa, control.circular=list(units="radians"))
```

The function **rvonmises** implements the acceptance–rejection approach of Best and Fisher (1979) based on the use of a wrapped Cauchy envelope. The value of the distribution function for $\theta = \pi/6$, $F(\pi/6) = 0.0543$, is computed using the command:

```
pvonmises(circular(pi/6), mu, kappa, from=circular(0))
```

and the value of the quantile function for $u = 0.5$, $Q(0.5) = 1.6139$, calculated using:

```
qvonmises(0.5, mu, kappa, from=circular(0), tol=.Machine$double.eps**(0.6))
```

The von Mises distribution was introduced by von Mises (1918) when studying the deviations of measured atomic weights from integral values. Mardia and Jupp (1999, pp. 41–43) describe five different constructions which lead to it. The first is a conditional distribution of the angular component of a polar representation of a bivariate normal distribution. The second is a maximum likelihood characterization analogous to that of Gauss for the normal distribution; namely that the von Mises distribution is the unique continuous circular distribution for which the mean direction, $\bar{\theta}$, is the maximum likelihood estimate of the population mean direction, μ. The third is as a maximum entropy distribution, and the last two arise from contexts involving diffusion processes on the circle and in the plane, respectively.

In many ways, the von Mises distribution plays an analogous role to that of the normal distribution for linear data. Indeed, in the literature it is often referred to as the *circular normal* distribution. Because of its appealing mathematical properties, techniques which assume circular data to be distributed according to it are far more developed than for any other circular distribution. As a consequence, and in parallel with the abuse of the normal distribution, those techniques have often been applied indiscriminately without due consideration of the assumptions underpinning them. We seek to partly address this problem by promoting the more flexible models described in Sections 4.3.9–4.3.13 of this chapter and the inferential techniques for them described in Chapter 6. As we shall see, those models are, or include, extensions of the von Mises distribution. In Section 6.2 we show how **R** can be used to fit the von Mises distribution to circular data.

4.3.9 *Jones–Pewsey Family*

The continuous circular uniform, cardioid, power-of-cosine, wrapped Cauchy, wrapped normal and von Mises distributions considered in the preceding sub-sections are the classical models of circular statistics. All of them, apart from the wrapped normal distribution, are actually special cases of a wider three-parameter family of reflectively symmetric circular distributions nowadays referred to as the Jones–Pewsey family. The probability density function of this family is

$$f(\theta) = \frac{\{\cosh(\kappa\psi) + \sinh(\kappa\psi)\cos(\theta - \mu)\}^{1/\psi}}{2\pi P_{1/\psi}(\cosh(\kappa\psi))}, \tag{4.55}$$

where μ is a location parameter, $\kappa \geq 0$ is a concentration parameter akin to that of the von Mises distribution, $-\infty < \psi < +\infty$ is a shape parameter and $P_{1/\psi}(z)$ is the *associated Legendre function* of the first kind of degree $1/\psi$ and order 0 (Gradshteyn and Ryzhik, 1994, Sections 8.7 and 8.8). The five classical models are obtained when: $\kappa = 0$ or $\psi \to \pm\infty$ and κ finite (continuous circular uniform); $\psi = 1$ (cardioid); $\psi > 0$ and $\kappa \to \infty$ (Cartwright's power-of-cosine); $\psi = -1$ (wrapped Cauchy); $\psi \to 0$ (von Mises).

Values of the normalizing constant of density (4.55) can be computed using the function **JPNCon** below which employs numerical integration when calculating

$$2\pi P_{1/\psi}(\cosh(\kappa\psi)) = \int_{-\pi}^{\pi} \{\cosh(\kappa\psi) + \sinh(\kappa\psi)\cos(\theta)\}^{1/\psi} d\theta. \tag{4.56}$$

```
JPNCon <- function(kappa, psi) {
if (kappa < 0.001) {ncon <- 1/(2*pi) ; return(ncon) }
else {
eps <- 10*.Machine$double.eps
if (abs(psi) <= eps) { ncon <- 1/(2*pi*I.0(kappa)) ; return(ncon) }
else {
intgrnd <- function(x) { (cosh(kappa*psi)+sinh(kappa*psi)*cos(x))**(1/psi) }
ncon <- 1/integrate(intgrnd, lower=-pi, upper=pi)$value
return(ncon) } }
}
```

Note the use within **JPNCon** of the **circular** library's **I.0** function to compute values of the zeroth order Bessel function of the first kind.

The polar representations of the densities portrayed in Fig. 4.6 exhibit many of the main features of Jones–Pewsey densities. For all but the continuous circular uniform case (for which the mean direction does not exist), the polar representation of the density is unimodal about μ, the mean direction. As κ decreases, the effect of ψ diminishes and, for a fixed value of κ, a density with $\psi \leq 0$ is more concentrated about μ and has lighter shoulders

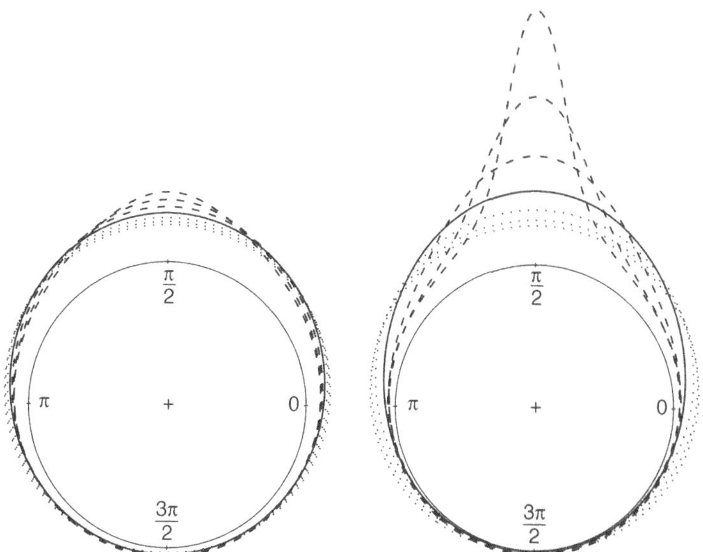

Figure 4.6 Polar representations of Jones–Pewsey densities with $\mu = \pi/2, \kappa = 1$ (left) and $\kappa = 2$ (right) and, in order of decreasing height at the mode, $\psi = -3/2, -1, -1/2$ (dashed; $\psi = -1$ is wrapped Cauchy), $\psi = 0$ (solid; von Mises) and $\psi = 1/2, 1, 3/2$ (dotted; $\psi = 1$ is cardioid)

than one with $\psi \geq 0$. When $\psi > 0$, there is relatively little difference between the densities of distributions with different κ-values.

Within **R**'s **circular** package, a polar representation of the Jones–Pewsey density with $\mu = \pi/2, \kappa = 2$ and $\psi = -3/2$ can be obtained using the commands:

```
mu <- circular(pi/2) ; kappa <- 2 ; psi <- -3/2
curve.circular(djonespewsey(x, mu, kappa, psi), join=TRUE, ylim=c(-0.9, 2.6), lwd=2, cex=0.8))
```

Note, however, that the values of the density returned by the **circular** library's **djonespewsey** function are incorrect if $\psi = 0$. Below we provide the function **JPPDF** for calculating values of the density (4.55) correctly.

For all but the continuous circular uniform case, the mean direction is μ and the distribution is reflectively symmetric about it. Thus, generally, $\bar{\beta}_p = 0$, and it can be shown that

$$\bar{\alpha}_p = \begin{cases} \dfrac{\Gamma((1/\psi)+1)P^p_{1/\psi}(\cosh(\kappa\psi))}{\Gamma((1/\psi)+p+1)P_{1/\psi}(\cosh(\kappa\psi))}, & \psi > 0, \\[2ex] I_p(\kappa)/I_0(\kappa), & \psi = 0, \\[2ex] \dfrac{\Gamma((1/|\psi|)-p)P^p_{1/\psi}(\cosh(\kappa\psi))}{\Gamma(1/|\psi|)P_{1/\psi}(\cosh(\kappa\psi))}, & \psi < 0. \end{cases} \tag{4.57}$$

Hence, $\tau_{p,\mu} = \bar{\alpha}_p$, and it follows, using (4.19), that $\tau_{p,0} = \bar{\alpha}_p e^{ip\mu}$. Thus, for all but the continuous circular uniform case,

$$\alpha_p = \bar{\alpha}_p \cos p\mu, \quad \beta_p = \bar{\alpha}_p \sin p\mu, \quad \rho_p = \bar{\alpha}_p, \quad \mu_p = p\mu.$$

In particular, the mean resultant length is $\rho = \bar{\alpha}_1$. Values of $\bar{\alpha}_p$ can be computed in **R** using the **JPNCon** function defined earlier and the **I.0**, **I.1** and **I.p** functions of the **circular** package.

There is no general closed-form expression for the distribution function and its values must be calculated using numerical integration of the density (4.55), i.e. as

$$F(\theta) = \frac{1}{2\pi P_{1/\psi}(\cosh(\kappa\psi))}\int_0^\theta \{\cosh(\kappa\psi) + \sinh(\kappa\psi)\cos(\phi-\mu)\}^{1/\psi}\,d\phi. \tag{4.58}$$

Values of the quantile function can be calculated using numerical inversion of the distribution function. There are no functions in **R**'s **circular** library to compute values of either function. Below we present functions with which to compute values of (4.55), (4.58) and the quantile function. For reasons of efficiency, particularly when simulating from a specified Jones–Pewsey distribution, all three functions have the normalizing constant **ncon**, calculated using the function **JPNCon**, as one of their arguments.

```
JPPDF <- function(theta, mu, kappa, psi, ncon) {
if (kappa < 0.001) {pdfval <- 1/(2*pi) ; return(pdfval)}
else {
eps <- 10*.Machine$double.eps
if (abs(psi) <= eps) {
```

```
pdfval <- ncon*exp(kappa*cos(theta-mu)) ; return(pdfval) }
else {
pdfval <- (cosh(kappa*psi)+sinh(kappa*psi)*cos(theta-mu))**(1/psi)
pdfval <- ncon*pdfval ; return(pdfval) } }
}

JPDF <- function(theta, mu, kappa, psi, ncon) {
eps <- 10*.Machine$double.eps
if (theta <= eps) { dfval <- 0 ; return(dfval) } else
if (theta >= 2*pi-eps) { dfval <- 1 ; return(dfval) } else
if (kappa < 0.001) {dfval <- theta/(2*pi) ; return(dfval)}
else {
if (abs(psi) <= eps) {
vMPDF <- function(x) { ncon*exp(kappa*cos(x-mu)) }
dfval <- integrate(vMPDF, lower=0, upper=theta)$value
return(dfval) }
else {
dfval <- integrate(JPPDF, mu=mu, kappa=kappa, psi=psi, ncon=ncon, lower=0, upper=theta)$value
return(dfval) } }
}

JPQF <- function(u, mu, kappa, psi, ncon) {
eps <- 10*.Machine$double.eps
if (u <= eps) {theta <- 0 ; return(theta)} else
if (u >= 1-eps) {theta <- 2*pi-eps ; return(theta)} else
if (kappa < 0.001) {theta <- u*2*pi ; return(theta)}
else {
roottol <- .Machine$double.eps**(0.6)
qzero <- function(x) { y <- JPDF(x, mu, kappa, psi, ncon)-u ; return(y) }
res <- uniroot(qzero, lower=0, upper=2*pi-eps, tol=roottol)
theta <- res$root ; return(theta) }
}
```

Now we have these three new functions at our disposal it is simple to calculate, for example, $F(\pi/2) = 0.4401445$ and $Q(0.4401445) = \pi/2$ for a Jones–Pewsey distribution with $\mu = \pi/2, \kappa = 2$ and $\psi = -3/2$, using the additional commands:

```
theta <- pi/6 ; mu <- pi/2 ; ncon <- JPNCon(kappa, psi)
dfjp <- JPDF(theta, mu, kappa, psi, ncon) ; dfjp
JPQF(dfjp, mu, kappa, psi, ncon)
```

The function **JPSim** below uses an acceptance–rejection algorithm with a rectangular envelope to simulate a random sample of size n from a Jones–Pewsey distribution with specified values of μ, κ and ψ and the corresponding value of the density's normalizing constant. For very concentrated cases of the distribution, the function **JPSim2** available from the website, which makes use of the **JPQF** function to implement inverse transform sampling, might be computationally more efficient.

```
JPSim <- function(n, mu, kappa, psi, ncon) {
fmax <- JPPDF(mu, mu, kappa, psi, ncon) ; theta <- 0
for (j in 1:n) { stopgo <- 0
while (stopgo == 0) {
u1 <- runif(1, 0, 2*pi) ; pdfu1 <- JPPDF(u1, mu, kappa, psi, ncon)
u2 <- runif(1, 0, fmax)
if (u2 <= pdfu1) { theta[j] <- u1 ; stopgo <- 1 }
} }
return(theta)
}
```

For instance, to simulate a random sample of size $n = 100$ from the Jones–Pewsey distribution with $\mu = \pi/2$, $\kappa = 2$ and $\psi = -3/2$ the only additional commands required are:

```
n <- 100 ; jpsamp <- JPSim(n, mu, kappa, psi, ncon)
```

Jones and Pewsey (2005) identify two constructions which lead to their family of distributions, based on conditioning spherically and elliptically symmetric distributions, respectively, onto the unit circle. In Section 6.3 we show how likelihood-based methods can be used to fit the Jones–Pewsey distribution to circular data.

4.3.10 Unimodal Symmetric Transformation of Argument Families

The three-parameter Jones–Pewsey family considered in the previous subsection is clearly more flexible than the individual classical symmetric circular distributions contained within it. Nevertheless, it does not contain distributions that are particularly *flat-topped*. As we have seen, the density (4.55) of the Jones–Pewsey family contains a $\cos(\theta - \mu)$ term. The same is true for the classical circular models contained within it. One approach to obtaining symmetric unimodal families containing distributions ranging from the relatively peaked to the relatively flat-topped is to apply a particular type of *transformation of argument* in which the $\cos(\theta - \mu)$ term is replaced by the function

$$\cos(\theta - \mu + \nu \sin(\theta - \mu)), \tag{4.59}$$

where ν is a *shape* parameter and, in order to ensure unimodality, $-1 < \nu < 1$. The original density is obtained when $\nu = 0$. Otherwise, the density is more flat-topped (peaked) than the original if $-1 < \nu < 0$ $(0 < \nu < 1)$.

Applying this approach to the von Mises distribution, the density of the resulting extended class of distributions is

$$f(\theta) = c\exp\{\kappa\cos(\theta - \mu + \nu\sin(\theta - \mu))\}, \tag{4.60}$$

where, as for the von Mises distribution, $\kappa \geq 0$ is the concentration parameter and, when $\kappa > 0$, μ is the mean direction. When $\nu \neq 0$, the normalizing constant,

$$c = \frac{1}{\int_{-\pi}^{\pi}\exp\{\kappa\cos(\theta + \nu\sin\theta)\}d\theta}, \tag{4.61}$$

must be recomputed numerically. Examples of density (4.60) with $\mu = \pi/2$, $\kappa = 3$ and a range of ν-values are portrayed in Fig. 4.7. There, the squashing-down and stretching-out effects of the shape parameter ν on the density around the mode are evident.

Throughout the remainder of this subsection we will consider computation in **R** for a distribution with density (4.60) and $\mu = \pi/2$, $\kappa = 2$ and $\nu = -1/2$. Below we provide four functions written to calculate values of the normalizing constant (4.61), density (4.60) and the corresponding distribution and quantile functions. Once more, the last two are based

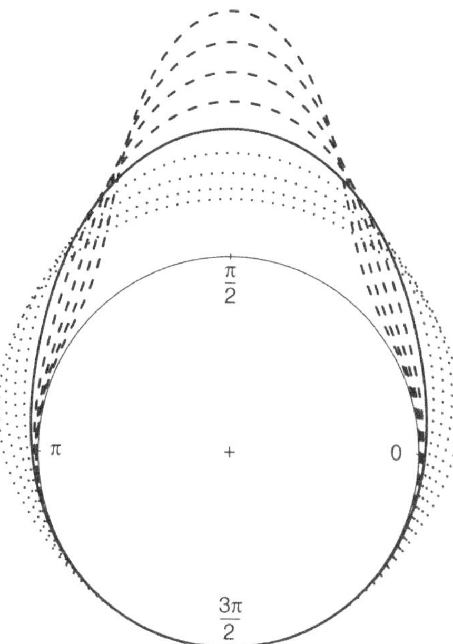

Figure 4.7 Polar representations of density (4.60) with $\mu = \pi/2$, $\kappa = 3$ and: $\nu = -0.999, -0.75, -0.5, -0.25$ (dotted); $\nu = 0$ (solid); $\nu = 0.25, 0.5, 0.75, 0.999$ (dashed). The solid line delimits the base von Mises density

on the use of numerical integration of the density and the probability integral transformation, respectively. As a means of improving efficiency, the last three functions have the normalizing constant **ncon**, calculated using the function **BatNCon**, as their last argument.

```
BatNCon <- function(kappa, nu) {
intgrnd <- function(x) { exp(kappa*cos(x+nu*sin(x))) }
ncon <- 1/(integrate(intgrnd, lower=-pi, upper=pi)$value)
return(ncon) }
}

BatPDF <- function(theta, mu, kappa, nu, ncon) {
pdfval <- ncon*exp(kappa*cos((theta-mu)+nu*sin(theta-mu)))
return(pdfval)
}

BatDF <- function(theta, mu, kappa, nu, ncon) {
eps <- 10*.Machine$double.eps
if (theta <= eps) { dfval <- 0 ; return(dfval) } else
if (theta >= 2*pi-eps) { dfval <- 1 ; return(dfval) }
else {
dfval <- integrate(BatPDF, mu=mu, kappa=kappa, nu=nu, ncon=ncon, lower=0, upper=theta)$value
return(dfval) }
}

BatQF <- function(u, mu, kappa, nu, ncon) {
eps <- 10*.Machine$double.eps
if (u <= eps) {theta <- 0 ; return(theta)} else
```

```
if (u >= 1-eps) {theta <- 2*pi-eps ; return(theta)}
else {
roottol <- .Machine$double.eps**(0.6)
qzero <- function(x) { y <- BatDF(x, mu, kappa, nu, ncon)-u ; return(y) }
res <- uniroot(qzero, lower=0, upper=2*pi-eps, tol=roottol)
theta <- res$root ; return(theta) }
}
```

With the aid of these new functions we can, for instance, produce a polar representation of the density of the distribution under consideration using the commands:

```
mu <- pi/2 ; kappa <- 2 ; nu <- -1/2 ; ncon <- BatNCon(kappa, nu)
curve.circular(BatPDF(x, mu, kappa, nu, ncon), join=TRUE, n=3600, ylim=c(-1, 2.1), lwd=2, cex=0.8)
```

As further illustrations of their use, we can compute $F(3\pi/4) = 0.7120$ and $Q(0.5) = 1.7301$ using the additional commands:

```
theta <- 3*pi/4 ; BatDF(theta, mu, kappa, nu, ncon)
BatQF(0.5, mu, kappa, nu, ncon)
```

The function **BatSim** below uses an acceptance–rejection algorithm to simulate a random sample of size n from a distribution with density (4.60) and specified values of μ, κ and ν and the corresponding value of the density's normalizing constant.

```
BatSim <- function(n, mu, kappa, nu, ncon) {
fmax <- BatPDF(mu, mu, kappa, nu, ncon) ; theta <- 0
for (j in 1:n) { stopgo <- 0
while (stopgo == 0) {
u1 <- runif(1, 0, 2*pi) ; pdfu1 <- BatPDF(u1, mu, kappa, nu, ncon)
u2 <- runif(1, 0, fmax)
if (u2 <= pdfu1) { theta[j] <- u1 ; stopgo <- 1 } } }
return(theta)
}
```

Then, to simulate a random sample of size $n = 100$ from the specific distribution under consideration, the only additional commands required are:

```
n <- 100 ; batsamp <- BatSim(n, mu, kappa, psi, ncon)
```

The class of symmetric extended von Mises distributions with density (4.60) was originally proposed by Batschelet (1981, Section 15.7), and studied more extensively by Pewsey *et al.* (2011). The class obtained by applying the same approach to the cardioid distribution was first investigated by Papakonstantinou (1979), and studied in considerably more depth by Abe *et al.* (2009). General properties of distributions derived using this transformation of argument approach are given in Abe *et al.* (2013), a paper in which the family of distributions obtained by applying the approach to the Jones–Pewsey family is also proposed and studied.

4.3.11 *Sine-skewed Distributions*

All of the models discussed so far are reflectively symmetric. Here we consider a general approach which can be used to obtain densities manifesting relatively low levels of asymmetry.

In Section 4.3.4 we saw how the cardioid density can be obtained by cosine perturbation of the continuous circular uniform density. Extending this idea, Umbach and Jammalamadaka (2009) adapt the perturbation approach of Azzalini (1985) to the circular context. One of the special cases of their general approach is the *sine-skewed* family of distributions studied by Abe and Pewsey (2011*a*). If $g(\theta - \xi)$ denotes the density of a reflectively symmetric circular base distribution with location parameter ξ then the density of its sine-skewed class is

$$f(\theta) = g(\theta - \xi)(1 + \lambda \sin(\theta - \xi)), \tag{4.62}$$

where $\lambda \in [-1, 1]$ is a skewing parameter. The symmetric base density is unperturbed when $\lambda = 0$, otherwise it is skewed in the counterclockwise direction ($\lambda > 0$) or the clockwise direction ($\lambda < 0$). Also, $f(\xi - \theta; \lambda) = f(\xi + \theta; -\lambda)$, and $f(\xi) = g(0)$ whatever the value of λ. Clearly, ξ will not generally be the mean direction of a sine-skewed distribution. Note also that densities obtained using this form of perturbation are, in fact, not necessarily unimodal. An appealing feature of this construction is that the normalizing constant is the same as that of the base density, and hence does not need to be recomputed.

Abe and Pewsey (2011*a*) propose the *sine-skewed Jones–Pewsey* family of distributions with density

$$f(\theta) = (1 + \lambda \sin(\theta - \xi))\frac{\{\cosh(\kappa\psi) + \sinh(\kappa\psi)\cos(\theta - \xi)\}^{1/\psi}}{2\pi P_{1/\psi}(\cosh(\kappa\psi))}, \tag{4.63}$$

as an asymmetric extension of the Jones–Pewsey distribution, and give the details of its fundamental properties. Polar representations of (4.63) for various combinations of ψ and λ are portrayed in Fig. 4.8.

There are no closed-form expressions for the distribution and quantile functions of the sine-skewed Jones–Pewsey distribution with density (4.63) and they must therefore be computed numerically. Below we provide functions for computing values of the density, distribution and quantile functions, and for simulating sine-skewed Jones–Pewsey variates. The first, **ssJPPDF**, calls the function **JPPDF** defined in Section 4.3.9. The fourth, **ssJPSim**, uses a basic acceptance–rejection algorithm. All four functions have **ncon**, the value of the normalizing constant of the base Jones–Pewsey distribution calculated using **JPNCon**, as their last argument.

```
ssJPPDF <- function(theta, xi, kappa, psi, lambda, ncon) {
pdfval <- JPPDF(theta, xi, kappa, psi, ncon)*(1+lambda*sin(theta-xi))
return(pdfval)
}
```

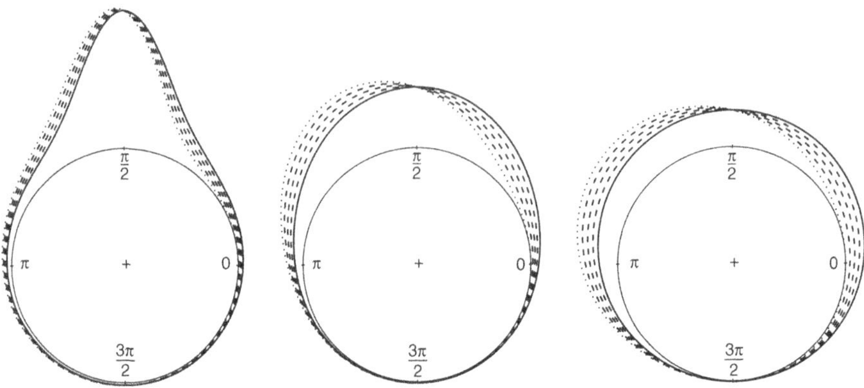

Figure 4.8 Polar representations of sine-skewed Jones–Pewsey densities with $\xi = \pi/2$, $\kappa = 2$ and: left $\psi = -1$ (sine-skewed wrapped Cauchy); centre $\psi = 0$ (sine-skewed von Mises); right $\psi = 1$ (sine-skewed cardioid). In each plot, the line types represent: $\lambda = 0$ (solid); $\lambda = 0.25, 0.5, 0.75$ (dashed); $\lambda = 1$ (dotted)

```
ssJPDF <- function(theta, xi, kappa, psi, lambda, ncon) {
eps <- 10*.Machine$double.eps
if (theta <= eps) { dfval <- 0 ; return(dfval) } else
if (theta >= 2*pi-eps) { dfval <- 1 ; return(dfval) }
else {
dfval <- integrate(ssJPPDF, xi=xi, kappa=kappa, psi=psi, lambda=lambda, ncon=ncon, lower=0,
upper=theta)$value
return(dfval) }
}

ssJPQF <- function(u, xi, kappa, psi, lambda, ncon) {
eps <- 10*.Machine$double.eps
if (u <= eps) {theta <- 0 ; return(theta)} else
if (u >= 1-eps) {theta <- 2*pi-eps ; return(theta)}
else {
roottol <- .Machine$double.eps**(0.6)
qzero <- function(x) {
y <- ssJPDF(x, xi, kappa, psi, lambda, ncon)-u
return(y) }
res <- uniroot(qzero, lower=0, upper=2*pi-eps, tol=roottol)
theta <- res$root ; return(theta) }
}

ssJPSim <- function(n, xi, kappa, psi, lambda, ncon) {
tval <- seq(0, 2*pi, by=2*pi/1440)
fval <- ssJPPDF(tval, xi, kappa, psi, lambda, ncon)
fmax <- max(fval) ; theta <- 0
for (j in 1:n) { stopgo <- 0
while (stopgo == 0) {
u1 <- runif(1, 0, 2*pi) ; pdfu1 <- ssJPPDF(u1, xi, kappa, psi, lambda, ncon)
u2 <- runif(1, 0, fmax)
if (u2 <= pdfu1) { theta[j] <- u1 ; stopgo <- 1 } } }
return(theta)
}
```

To illustrate the use of the four new functions we employ them for the sine-skewed Jones–Pewsey density with $\xi = \pi/2$, $\kappa = 2$, $\psi = 1$ and $\lambda = 0.5$. A polar representation of its density can be plotted using the commands:

```
xi <- pi/2 ; kappa <- 2 ; psi <- 1 ; lambda <- 0.5
ncon <- JPNCon(kappa, psi)
curve.circular(ssJPPDF(x, xi, kappa, psi, lambda, ncon), join=TRUE, n=3600,
ylim=c(-1,2.1), lwd=2, cex=0.8)
```

For the same distribution, the values of $F(3\pi/2) = 0.9446$ and $Q(0.5) = 2.1879$ can be computed using the commands:

```
ssJPDF(3*pi/2, xi, kappa, psi, lambda, ncon)
ssJPQF(0.5, xi, kappa, psi, lambda, ncon)
```

and a sample of size $n = 100$ simulated from it using the commands:

```
n <- 100 ; ssjpsamp <- ssJPSim(n, xi, kappa, psi, lambda, ncon)
```

4.3.12 *Unimodal Asymmetric Transformation of Argument Families*

We first saw the application of the transformation of argument approach in Section 4.3.10 where the $\cos(\theta - \mu)$ term appearing in a circular density was replaced by (4.59). By so doing, the range of densities available in a base distribution was widened to include ones which were more peaked and others which were more flat-topped. Here we consider a slight adaptation of that form of transformation of argument, designed to skew a base symmetric density containing the term $\cos(\theta - \mu)$. It provides an alternative means of producing asymmetric distributions, with wider ranges of asymmetry than the sine-skewed models considered in the previous subsection and with unimodality guaranteed.

Rather than substituting (4.59) for $\cos(\theta - \mu)$ in the base density, here we consider densities generated by replacing it by

$$\cos(\theta - \xi + \nu \cos(\theta - \xi)), \tag{4.64}$$

where ξ is a location parameter and, now, ν is an *asymmetry* parameter. Once again, in order to ensure unimodality, $-1 < \nu < 1$. The original density is obtained when $\nu = 0$. Otherwise, the density is skewed counterclockwise (clockwise) if $0 < \nu < 1$ $(-1 < \nu < 0)$. When $\nu \neq 0$ the normalizing constant of the skewed distribution must generally be recomputed using numerical integration. Only when $\nu = 0$ does ξ represent the mean direction.

This approach was first applied to the cardioid distribution by Papakonstantinou (1979), and then to the von Mises distribution in Batschelet (1981, Section 15.6). Abe *et al.* (2013) study the general properties of distributions obtained using this construction and propose an asymmetric extended Jones–Pewsey family of distributions with density

$$f(\theta) = c\{\cosh(\kappa\psi) + \sinh(\kappa\psi)\cos(\theta - \xi + \nu\cos(\theta - \xi))\}^{1/\psi}, \tag{4.65}$$

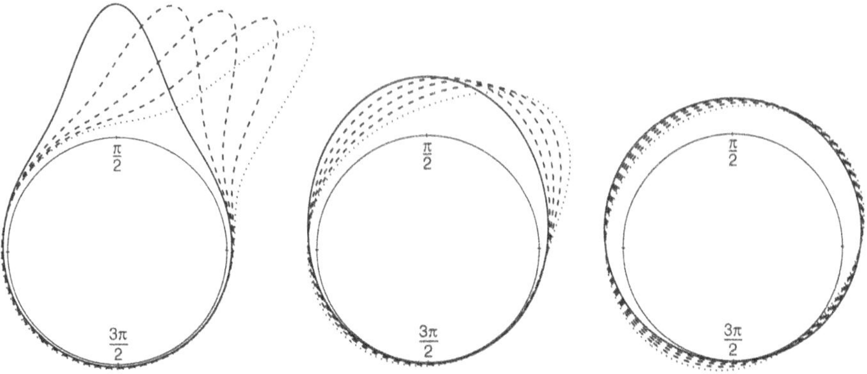

Figure 4.9 Polar representations of asymmetric extended Jones–Pewsey densities with $\xi = \pi/2$, $\kappa = 2$ and: left $\psi = -1$ (asymmetric extended wrapped Cauchy); centre $\psi = 0$ (asymmetric extended von Mises); right $\psi = 1$ (asymmetric extended cardioid). In each plot, the line types represent: $\nu = 0$ (solid); $\nu = 0.25, 0.5, 0.75$ (dashed); $\nu = 0.999$ (dotted)

where the normalizing constant,

$$c = \frac{1}{\int_{-\pi}^{\pi} \{\cosh(\kappa\psi) + \sinh(\kappa\psi)\cos(\theta + \nu\cos\theta)\}^{1/\psi} d\theta}, \tag{4.66}$$

must be recomputed numerically when $\nu \neq 0$. Examples of density (4.65) with $\xi = \pi/2$, $\kappa = 2$, $\psi = -1, 0, 1$ and various positive ν-values are presented in Fig. 4.9. The skewing effect of ν is clear from this plot. Densities for negative ν-values are simply the reflections about ξ of their positive ν-valued counterparts. The mode of the density is at $\xi + H_\nu^{-1}(\pi/2) - \pi/2 \pmod{2\pi}$, where $H_\nu(\theta) = \theta + \nu \sin\theta$ must be inverted numerically.

There are no closed-form expressions for the distribution and quantile functions of the asymmetric extended Jones–Pewsey distribution with density (4.65) and they must be computed using numerical integration and root-finding, respectively. Below we present functions to calculate values of the normalizing constant (4.66), density (4.65), distribution function and quantile function, respectively. The fifth function, **aeJPSim**, uses an acceptance–rejection algorithm to simulate asymmetric extended Jones–Pewsey random variates. The last four functions have **ncon**, the value of the normalizing constant of the distribution calculated using **aeJPNCon**, as their last argument.

```
aeJPNCon <- function(kappa, psi, nu) {
eps <- 10*.Machine$double.eps
if (abs(psi) <= eps) {
intgrnd0 <- function(x) { exp(kappa*cos(x+nu*cos(x))) }
ncon <- 1/(integrate(intgrnd0, lower=-pi, upper=pi)$value) ; return(ncon) }
else {
intgrnd <- function(x) {
(cosh(kappa*psi)+sinh(kappa*psi)*cos(x+nu*cos(x)))**(1/psi)}
ncon <- 1/(integrate(intgrnd, lower=-pi, upper=pi)$value) ; return(ncon) }
}
```

```
aeJPPDF <- function(theta, xi, kappa, psi, nu, ncon) {
eps <- 10*.Machine$double.eps
if (abs(psi) <= eps) {
pdfval <- ncon*exp(kappa*cos(theta-xi+nu*cos(theta-xi))) ; return(pdfval) }
else {
pdfval <- (cosh(kappa*psi)+sinh(kappa*psi)*cos(theta-xi+nu*cos(theta-xi)))**(1/psi)
pdfval <- ncon*pdfval
return(pdfval) }
}

aeJPDF <- function(theta, xi, kappa, psi, nu, ncon) {
eps <- 10*.Machine$double.eps
if (theta <= eps) { dfval <- 0 ; return(dfval) } else
if (theta >= 2*pi-eps) { dfval <- 1 ; return(dfval) }
else {
dfval <- integrate(aeJPPDF, xi=xi, kappa=kappa, psi=psi, nu=nu, ncon=ncon,
lower=0, upper=theta)$value
return(dfval) }
}

aeJPQF <- function(u, xi, kappa, psi, nu, ncon) {
eps <- 10*.Machine$double.eps
if (u <= eps) {theta <- 0 ; return(theta)} else
if (u >= 1-eps) {theta <- 2*pi-eps ; return(theta)}
else {
roottol <- .Machine$double.eps**(0.6)
qzero <- function(x) { y <- aeJPDF(x, xi, kappa, psi, nu, ncon)-u
return(y) }
res <- uniroot(qzero, lower=0, upper=2*pi-eps, tol=roottol)
theta <- res$root ; return(theta) }
}

aeJPSim <- function(n, xi, kappa, psi, nu, ncon) {
eps <- 10*.Machine$double.eps ; roottol <- .Machine$double.eps**(0.6)
qzero <- function(x) { y <- x+nu*sin(x)-pi/2 ; return(y) }
res <- uniroot(qzero, lower=0, upper=2*pi-eps, tol=roottol)
mode <- xi+res$root-pi/2
fmax <- aeJPPDF(mode, xi, kappa, psi, nu, ncon) ; theta <- 0
for (j in 1:n) { stopgo <- 0
while (stopgo == 0) {
u1 <- runif(1, 0, 2*pi) ; pdfu1 <- aeJPPDF(u1, xi, kappa, psi, nu, ncon)
u2 <- runif(1, 0, fmax)
if (u2 <= pdfu1) { theta[j] <- u1 ; stopgo <- 1 } } }
return(theta)
}
```

To illustrate the application of these five new functions we use them for the particular case of the asymmetric extended Jones–Pewsey density with $\xi = \pi/2$, $\kappa = 2$, $\psi = -1$ and $\nu = 0.75$. A polar representation of its density can be produced using the commands:

```
xi <- pi/2 ; kappa <- 2 ; psi <- -1 ; nu <- 0.75
ncon <- aeJPNCon(kappa, psi, nu)
curve.circular(aeJPPDF(x, xi, kappa, psi, nu, ncon), join=TRUE, n=3600, ylim=c(-1,2.1), lwd=2,
cex=0.8)
```

and its values of $F(\pi/2) = 0.7535$ and $Q(0.5) = 1.0499$ evaluated using the commands:

```
aeJPDF(pi/2, xi, kappa, psi, nu, ncon) ; aeJPQF(0.5, xi, kappa, psi, nu, ncon)
```

A random sample of size $n = 100$ can be simulated from it using the commands:

```
n <- 100 ; aejpsamp <- aeJPSim(n, xi, kappa, psi, nu, ncon)
```

4.3.13 *Inverse Batschelet Distributions*

Inspired by the transformation of argument approaches considered in Sections 4.3.10 and 4.3.12, Jones and Pewsey (2012) propose *inverse Batschelet* distributions generated using the related idea of *transformation of scale*. Their approach involves the combination of the inverse of the function

$$t_{1,\nu}(\theta) = \theta - \nu(1 + \cos\theta), \tag{4.67}$$

and the function

$$t_\lambda(\theta) = \begin{cases} \frac{1-\lambda}{1+\lambda}\theta + \frac{2\lambda}{1+\lambda}s_\lambda^{-1}(\theta), & -1 < \lambda \leq 1, \\ \theta - \sin\theta, & \lambda = -1, \end{cases} \tag{4.68}$$

where

$$s_\lambda(\theta) = \theta - \frac{1}{2}(1 + \lambda)\sin\theta, \tag{4.69}$$

$-1 \leq \nu \leq 1$ is a skewness parameter and $-1 \leq \lambda \leq 1$ is a parameter which regulates the peakedness of the distribution. The resulting four-parameter distributions are unimodal and display the widest ranges of both skewness and peakedness yet available. Applying their approach to a base von Mises distribution results in a unimodal family of distributions with density

$$f(\theta) = \frac{1}{2\pi I_0(\kappa)K_{\kappa,\lambda}} \exp\left\{\kappa \cos\left[\frac{1-\lambda}{1+\lambda}t_\nu(\theta - \xi) + \frac{2\lambda}{1+\lambda}s_\lambda^{-1}(t_\nu(\theta - \xi))\right]\right\}, \tag{4.70}$$

where $t_\nu(\theta) = t_{1,\nu}^{-1}(\theta)$, ξ is a location parameter, $\kappa \geq 0$ is the concentration parameter,

$$K_{\kappa,\lambda} = \begin{cases} \frac{1+\lambda}{1-\lambda} - \frac{2\lambda}{(1-\lambda)2\pi I_0(\kappa)} \int_{-\pi}^{\pi} \exp[\kappa \cos\{\theta - \frac{1}{2}(1-\lambda)\sin\theta\}]d\theta, & -1 \leq \lambda < 1, \\ 1 - A_1(\kappa), & \lambda = 1, \end{cases} \tag{4.71}$$

and $I_0(\kappa)$ and $A_1(\kappa)$ are as defined in (4.52) and (4.53), respectively. The inversion of the functions $t_{1,\nu}$ and s_λ, as well as the computation of the constant $K_{\kappa,\lambda}$, must generally be performed numerically. When $\nu = 0$ and $\kappa > 0$, the mean direction is ξ (mod 2π). More generally, when $\kappa > 0$, the mode is $\xi - 2\nu$ (mod 2π). These two facts, as well as the roles of ν and κ, are illustrated by the densities portrayed in Fig. 4.10. The density for a negative ν-value is simply the reflection about ξ of the density with the corresponding positive ν-value.

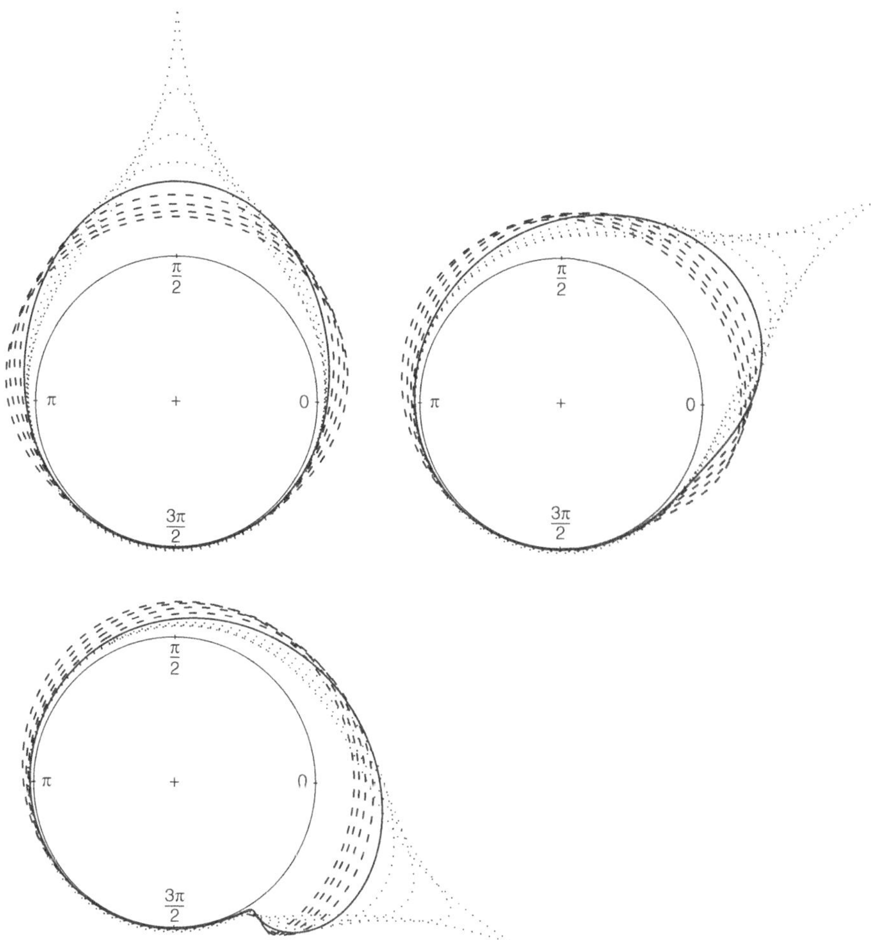

Figure 4.10 Polar representations of inverse Batschelet densities for a base von Mises distribution and $\xi = \pi/2$, $\kappa = 2$ and: top left, $\nu = 0$ (symmetric); top right, $\nu = 0.5$; bottom, $\nu = 1$. Corresponding to increasing height at the mode, the line types represent densities with: $\lambda = -1, -0.75, -0.5, -0.25$ (dashed); $\lambda = 0$ (solid); $\lambda = 0.25, 0.5, 0.75, 1$ (dotted)

Below we provide the basic functions **tnu, invslambda** and **invBNCon** written to compute values of the two inverse functions, t_ν and s_λ^{-1}, and the normalizing constant $1/(2\pi I_0(\kappa)K_{\kappa,\lambda})$.

```
tnu <- function(theta, xi, nu) {
phi <- theta-xi
if (phi <= -pi) {phi <- phi+2*pi}
else if (phi > pi) {phi <- phi-2*pi}
eps <- 10*.Machine$double.eps
```

```
if (phi <= -pi+eps) {return(-pi)}
else if (phi >= pi-eps) {return(pi)}
else {
roottol <- .Machine$double.eps**(0.6)
t1nuzero <- function(x) {
y <- x-nu*(1+cos(x))-phi ; return(y) }
res <- uniroot(t1nuzero, lower=-pi+eps, upper=pi, tol=roottol)
return(res$root) }
}

invslambda <- function(theta, lambda) {
eps <- 10*.Machine$double.eps
if (lambda <= -1+eps) {return(theta)}
else {
roottol <- .Machine$double.eps**(0.6)
islzero <- function(x) {
y <- x-(1+lambda)*sin(x)/2-theta ; return(y) }
res <- uniroot(islzero, lower=-pi+eps, upper=pi, tol=roottol)
return(res$root) }
}

invBNCon <- function(kappa, lambda) {
eps <- 10 * .Machine$double.eps
mult <- 2*pi*I.0(kappa)
if (lambda >= 1-eps) { ncon <- 1/(mult*(1-A1(kappa))) ; return(ncon) }
else {
con1 <- (1+lambda)/(1-lambda) ; con2 <- (2*lambda)/((1-lambda)*mult)
intgrnd <- function(x) { exp(kappa*cos(x-(1-lambda)*sin(x)/2)) }
intval <- integrate(intgrnd, lower=-pi, upper=pi)$value
ncon <- 1/(mult*(con1-con2*intval)) ; return(ncon) }
}
```

These three new functions are used within the functions **invBPDF**, **invBDF**, **invBQF** and **invBSim** below. The first three have been written to compute values of the density (4.70) and its corresponding distribution and quantile functions, respectively. The function **invBDF** employs a simple midpoint approximation to perform numerical integration of the density. In **invBSim**, an acceptance–rejection algorithm is used to simulate random variates from (4.70). For certain distributions, the algorithms proposed in Jones and Pewsey (2012) will be more efficient. All four functions have **ncon**, the value of the normalizing constant of the distribution, as their last argument.

```
invBPDF <- function(theta, xi, kappa, nu, lambda, ncon) {
arg1 <- tnu(theta, xi, nu) ; eps <- 10*.Machine$double.eps
if (lambda <= -1+eps) {
pdfval <- ncon*exp(kappa*cos(arg1-sin(arg1))) ; return(pdfval) }
else {
con1 <- (1-lambda)/(1+lambda) ; con2 <- (2*lambda)/(1+lambda)
arg2 <- invslambda(arg1, lambda)
pdfval <- ncon*exp(kappa*cos(con1*arg1+con2*arg2))
return(pdfval) }
}

invBDF <- function(theta, xi, kappa, nu, lambda, ncon) {
eps <- 10*.Machine$double.eps
if (theta <= eps) {dfval <- 0 ; return(dfval)} else
if (theta >= 2*pi-eps) {dfval <- 1 ; return(dfval)} else
if (theta <= pi/2) {nint <- 90} else
```

```
if (theta <= pi) {nint <- 180} else
if (theta <= 3*pi/2) {nint <- 270} else
if (theta < 2*pi-eps) {nint <- 360}
width <- theta/nint ; dfval <- 0
for (j in 1:nint) {
arg <- (2*j-1)*width/2
dfval <- dfval+invBPDF(arg, xi, kappa, nu, lambda, ncon) }
dfval <- dfval*width ; return(dfval)
}

invBQF <- function(u, xi, kappa, nu, lambda, ncon) {
eps <- 10*.Machine$double.eps
if (u <= eps) {theta <- 0 ; return(theta)} else
if (u >= 1-eps) {theta <- 2*pi-eps ; return(theta)}
else {
roottol <- 1000*.Machine$double.eps**(0.6)
qzero <- function(x) {
y <- invBDF(x, xi, kappa, nu, lambda, ncon)-u ; return(y) }
res <- uniroot(qzero, lower=0, upper=2*pi-eps, tol=roottol)
theta <- res$root ; return(theta) }
}

invBSim <- function(n, xi, kappa, nu, lambda, ncon) {
mode <- xi-2*nu ; fmax <- invBPDF(mode, xi, kappa, nu, lambda, ncon)
theta <- 0
for (j in 1:n) { stopgo <- 0
while (stopgo == 0) {
u1 <- runif(1, 0, 2*pi) ; pdfu1 <- invBPDF(u1, xi, kappa, nu, lambda, ncon)
u2 <- runif(1, 0, fmax)
if (u2 <= pdfu1) { theta[j] <- u1 ; stopgo <- 1 }
}}
return(theta)
}
```

To illustrate the use of these new functions, we apply them in calculations for an inverse Batschelet distribution with density (4.70), $\xi = \pi/2$, $\kappa = 2$, $\nu = -0.5$ and $\lambda = 0.7$. A polar representation of its density can be plotted using:

```
xi <- pi/2 ; kappa <- 2 ; nu <- -0.5 ; lambda <- 0.7
ncon <- invBNCon(kappa, lambda)
x <- seq(0, 2*pi, by=2*pi/3600) ; y <- 0
for (j in 1:3601) { y[j] <- invBPDF(x[j], xi, kappa, nu, lambda, ncon) }
theta <- circular(x)
curve.circular(dcircularuniform, join=TRUE, ylim=c(-1.7, 1), lty=0)
lines(theta, y, lwd=2)
```

The values of $F(\pi/2) = 0.1180$ and $Q(0.5) = 2.5138$ can be calculated using the commands:

```
invBDF(pi/2, xi, kappa, nu, lambda, ncon)
invBQF(0.5, xi, kappa, nu, lambda, ncon)
```

Finally, a sample of size $n = 100$ can be simulated using the commands:

```
n <- 100 ; invbsamp <- invBSim(n, xi, kappa, nu, lambda)
```

In Section 6.4 we show how likelihood-based methods can be used to fit the inverse Batschelet distribution with density (4.70).

4.3.14 *Summary of Continuous Circular Models*

As an aid to their assimilation, below we provide a brief summary of the main features of the continuous circular distributions described in Sections 4.3.3–4.3.13.

Circular uniform Characterizes circular randomness or isotropy, in which no direction is any more likely than any other. Reflectively symmetric about any axis. A limiting case of most circular models, and the limiting distribution of the circular equivalent of the central limit theorem.

Cardioid Arises from cosine perturbation of the continuous circular uniform density. Density has a closed form and is heart-like in shape. Unimodal and reflectively symmetric about the mean direction. A special case of the Jones–Pewsey family.

Cartwright's Power-of-cosine distribution related to the cardioid. Density has a closed form, is unimodal and reflectively symmetric about the mean direction. A special case of the Jones–Pewsey family.

Wrapped Cauchy Obtained by wrapping a Cauchy random variable around the circumference of the unit circle. Only known wrapped distribution for which the density has a closed form. Unimodal and reflectively symmetric about the mean direction. A special case of the Jones–Pewsey family. Simulation simple.

Wrapped normal Derived by wrapping a normal random variable around the circumference of the unit circle. Density involves an infinite sum which is generally well-approximated using relatively few of its central terms. Unimodal and reflectively symmetric about the mean direction. For a given mean resultant length, is less concentrated about the mean direction than the wrapped Cauchy. Provides a close approximation to the von Mises distribution. Simulation simple.

Von Mises The classic model of circular statistics. Plays an analogous role to that of the normal distribution for linear data. Normalizing constant involves a modified Bessel function. Unimodal and reflectively symmetric about the mean direction. A limiting case of the Jones–Pewsey family. Simulation involves the use of a wrapped Cauchy envelope.

Jones–Pewsey family Over-arching family of distributions which includes all of the above, apart from the wrapped normal, as special or limiting cases. Normalizing constant involves an associated Legendre polynomial. Arises from conditioning spherically and elliptically symmetric distributions. Unimodal and reflectively symmetric about the mean direction. Includes relatively peaked distributions but not ones that are very flat-topped.

Unimodal symmetric transformation of argument families Obtained by transforming the argument of the cosine term of a base unimodal reflectively symmetric density. Normalizing constant must be recomputed. Resulting densities range from the relatively peaked to the very flat-topped.

Sine-skewed Derived by perturbing a base unimodal reflectively symmetric density. Normalizing constant is the same as that of the base density. Resulting densities can model moderate departures from symmetry but are not guaranteed to be unimodal.

Unimodal asymmetric transformation of argument families Obtained by transforming the argument of the cosine term of a base unimodal reflectively symmetric density. Normalizing constant must be recomputed. Resulting densities are always unimodal and allow for substantial departures from symmetry.

Inverse Batschelet family The most flexible circular model to date. Can be used to model wide-ranging levels of asymmetry and peakedness. Based on the use of two inverse functions which must be computed numerically.

4.3.15 Other Models for Unimodal Data

In Sections 4.3.6 and 4.3.7 we considered the wrapped Cauchy and wrapped normal distributions, perhaps the two best known classical wrapped distributions. However, as explained in Section 4.3.1, the same wrapping approach can be applied to any linear distribution. In recent years, various alternative models obtained using the wrapping construction have been proposed in the literature. Amongst them, Pewsey *et al.* (2007) considered the *wrapped t* class of unimodal symmetric distributions. This class contains the wrapped Cauchy and wrapped normal distributions as special cases. Included amongst the unimodal asymmetric models that have been proposed recently are the *wrapped skew-normal* class of Pewsey (2000, 2006), the *wrapped exponential* and *wrapped Laplace* distributions of Jammalamadaka and Kozubowski (2004) (see also Pewsey (2002*a*)), and the *wrapped normal-Laplace* distributions of Reed and Pewsey (2009).

There are no functions available within the **circular** package associated with any of these more recently proposed wrapped distributions. However, it is not difficult to write **R** code, along similar lines to the code presented in the preceding subsections, to compute values of their density, distribution and quantile functions, and perform simulation from them. The one complication is to ensure that sufficient terms of the series (4.33) are included so as to obtain an adequate approximation of the density. For densities requiring the inclusion of many terms, their computation, and that of other functions related to them, can be very slow (see Pewsey *et al.* (2007)).

There is one other class of wrapped distributions that is partially supported within the **circular** package, namely the *wrapped stable* class (see Mardia (1972, page 57), Jammalamadaka and SenGupta (2001, Section 2.2.8) and Pewsey (2008)). However, the only function available, **rwrappedstable**, is for simulating wrapped stable random variates. The wrapped stable class also contains the wrapped Cauchy and wrapped normal distributions, as well as other reflectively symmetric distributions that are at least as peaked as the latter and others that are skew. As Pewsey (2008) shows, method of moments estimation for the wrapped stable class is surprisingly simple.

Finally, we mention the *exponential models* of Beran (1979) and the model of Kato and Jones (2010) derived using *Möbius transformation* of the von Mises distribution. Both can model varying levels of peakedness and asymmetry but the distributions contained within them are not necessarily unimodal. Moreover, their four-parameter unimodal distributions are nowhere near as flexible as the inverse Batschelet distributions considered in Section 4.3.13.

4.3.16 *Multimodal Models*

We have already mentioned various models with polar representations of their densities which can assume bimodal forms. Nevertheless, their bimodality is generally a by-product of constructions employed to induce asymmetry. The obvious general approach to modelling multimodal circular data is to use finite mixtures of any of the unimodal models considered in the previous subsections. One particularly appealing feature of mixture distributions is that their parameters are generally easy to interpret. Finite mixtures of von Mises distributions are by far the most thoroughly investigated and can also be used to model skew unimodal circular data (see Mardia and Sutton (1975), Spurr (1981), Bartels (1984) and Spurr and Koutbeiy (1991)). The **circular** package contains functions for computing values of the density and distribution functions of, and simulating data from, two-component von Mises mixture distributions, while the **movMF** package has functions to compute the density of, simulate data from, and fit, von Mises mixtures with two or more components. An alternative approach to modelling data with two diametrically opposed modes is that of Abe and Pewsey (2011*b*) based on duplication and cosine perturbation.

Another way of extending the von Mises model to produce asymmetric as well as multimodal distributions is to apply the notion of "multiplicative mixing" underpinning the so-called *generalized von Mises* distributions. These distributions have an extensive history, some of the most relevant references to them being Maksimov (1967), Ruhkin (1972), Cox (1975), Beran (1979), Yfantis and Borgman (1982), and Gatto and Jammalamadaka (2007). The density for the two-component, GvM_2, case is

$$f(\theta) = \frac{1}{2\pi\, G_0(\delta, \kappa_1, \kappa_2)} \exp\{\kappa_1 \cos(\theta - \mu_1) + \kappa_2 \cos 2(\theta - \mu_2)\}, \tag{4.72}$$

where μ_1 and μ_2 are location parameters, $\kappa_1, \kappa_2 \geq 0$ are concentration parameters and the constant $G_0(\delta, \kappa_1, \kappa_2)$ must generally be computed numerically. The calculation of the analogous constant for the obvious k-component extension, GvM_k, can be difficult. In the **circular** package the function **dgenvonmises** is available for computing values of (4.72), but otherwise generalized von Mises distributions are not supported.

Finally, we make brief reference to the non-negative trigonometric moment (NNTS) distributions of Fernández-Durán (2004). Whilst such distributions are well supported by **R**'s **CircNNTSR** package, we are unable to recommend their use. Firstly because, when fitted to real data, densities of this type generally involve many more parameters than those of the distributions that we have considered previously. Moreover, the interpretation of their parameters is far from simple. Also, because their densities are finite Fourier series constrained to be non-negative, fitted NNTS distributions tend to display minor harmonic modes that need not be supported by the data. Whilst NNTS distributions might provide a not too parameter-heavy fit to data drawn from a cardioid-like distribution, vast numbers of terms are required before a reasonable approximation to the cusp-like or very flat-topped cases of the inverse Batschelet distribution, for instance, are obtained. Neither have we come across any examples where NNTS distributions provide a convincing fit to real circular data.

Far better approximations to underlying densities can be obtained using suitably smoothed examples of the nonparametric kernel density estimates of Section 2.4.

4.3.17 *Models for Toroidal Data*

There are situations in which pairs of circular random variables, (Θ, Ψ), will be of interest. For instance, you might be interested in modelling the direction of the wind at a particular wind farm at 6am and again at 6pm. Or you could be interested in the joint distribution of the direction of flight of migrating birds and the direction of the prevailing wind. Pairs of random variables of this type are referred to as being *bivariate circular* or *toroidal* because, as they involve two circular random variables, their support, i.e. the geometrical structure upon which they live, is the unit torus. Doughnuts and life-buoys are (approximately) toroidal in shape. The distributions of pairs of circular random variables are also referred to as being toroidal or bivariate circular, as are the data obtained when the values taken by the pair of random variables are observed.

The most basic model for toroidal data is the bivariate continuous uniform distribution on the torus with density

$$f(\theta, \psi) = \frac{1}{4\pi^2},$$

corresponding to Θ and Ψ being two independent continuous circular uniform random variables. A bivariate von Mises distribution was proposed by Mardia (1975). More generally, Johnson and Wehrly (1977) proposed models with bivariate circular densities of the form

$$f(\theta, \psi) = 2\pi f_\Theta(\theta) f_\Psi(\psi) g(2\pi\{F_\Theta(\theta) \pm F_\Psi(\psi)\}), \tag{4.73}$$

where f_Θ and f_Ψ are the marginal densities of Θ and Ψ, F_Θ and F_Ψ are their marginal distribution functions, and g is also a circular density. Jones *et al.* (2013) study the properties of models with density (4.73) and, so as to stress the fact that they are the circular analogues of copulas, refer to them as *circulas*. A special case of such a distribution, derived using Möbius transformation and having wrapped Cauchy marginal as well as conditional distributions, is studied by Kato and Pewsey (2013).

Correlation and regression methods for use with bivariate circular data will be considered in Chapter 8.

4.3.18 *Models for Cylindrical Data*

Returning to the wind farm example of the previous subsection, suppose now that we are interested in the joint distribution of the wind's velocity and direction. More generally, we can characterize such situations as involving a random vector (X, Θ), where X denotes a linear random variable and Θ a circular one. The support for random vectors of this type is a cylinder with radius 1. The length of the cylinder depends on the support for X. It would be infinite, for example, if the marginal distribution of X were Gaussian. Random variables of this type, as well as their distributions and data collected on them, are usually referred to as being *cylindrical*.

Mardia and Sutton (1978) proposed a model for cylindrical data in which the distribution of Θ is von Mises and the conditional distribution of $X|\theta$ is normal. More generally, and using an analogous structure to that of the density in (4.73), Johnson and Wehrly (1978) proposed cylindrical models with densities of the form

$$f(\theta, x) = 2\pi f_\Theta(\theta) f_X(x) g(2\pi\{F_\Theta(\theta) - F_X(x)\}), \tag{4.74}$$

where f_X and F_X are the marginal density and distribution function of X and the other terms are as defined for (4.73). For instance, the marginal distributions of X and Θ might be normal and von Mises, respectively, and g a cardioid density.

We will also consider correlation and regression methods for use with cylindrical data in Chapter 8.

5

Basic Inference for a Single Sample

There are various issues of basic interest during the initial inferential phase of an exploratory analysis of circular data. Firstly, the null hypotheses of circular uniformity and reflective symmetry are fundamental dividing hypotheses in the sense that the rejection or non-rejection of either can lead us down very different subsequent inferential routes. Should uniformity not be rejected then there would be no need to look for a more complicated model for our data, whereas its rejection would require us to embark on such a search. The non-rejection of symmetry might well lead us to consider fitting the Jones-Pewsey family of distributions of Section 4.3.9 and the application of model reduction techniques in an attempt to identify a parsimonious model for our data. On the other hand, the rejection of symmetry would suggest that the fit of some flexible asymmetric family, such as the inverse Batschelet distributions considered in Section 4.3.13, should be explored. We will see how to fit these two models in Chapter 6. Testing for circular uniformity and reflective symmetry is considered in Sections 5.1 and 5.2.

Once isotropy and reflective symmetry have been investigated, we will often be interested in carrying out distribution-free point and interval estimation for key circular summaries such as the mean direction, mean resultant length and the second central sine and cosine moments. We will consider approaches to performing both forms of inference in Sections 5.3.1 and 5.3.2. Finally, in certain contexts it will be of interest to test for a specified mean direction, perhaps during the quality control of some device. Approaches to this type of inference are discussed in Section 5.3.3.

Where appropriate, we consider large-sample and bootstrap approaches to performing inference. As explained by Fisher (1993, Section 8.2), for very small-sized samples, it is possible to compute the complete resampling distribution of any estimator or test statistic. Alternatively, iterated bootstrap calibration can be used (see Fisher 1993, Section 8.3.3). However, for samples of such small size, the quality of any inferences drawn from them should be treated with caution, and, whenever possible, more data should be sought.

5.1 Testing for Uniformity

As we saw in Section 4.3.3, circular uniformity, or isotropy, is the most basic dividing hypothesis in circular statistics. If it cannot be rejected then the population from which the sample was drawn can reasonably be assumed to be evenly distributed around the circle. When uniformity is rejected the data support some form of departure from evenness.

Consider the data shown in Fig. 5.1 which represent the arrival times on the 24-hour clock of 254 patients at an intensive care unit, recorded over a period of around 12 months (see Fisher (1993, page 239) and Cox and Lewis (1966, page 254)). They are available as the circular data object **fisherB1c** within **R**'s **circular** package. Figure 5.1 was produced using the commands:

```
library(circular)
plot(fisherB1c, shrink=1.2, stack=TRUE, pch=16, bins=720, cex=1.5)
lines(density.circular(fisherB1c, bw=40), lwd=2)
rose.diag(fisherB1c, bins=24, cex=1.5, prop=2.3, col="grey", add=TRUE)
```

Visually it would appear that admissions are less common between around 3am and 9am, but how might we test this formally? The key issue here is to identify the alternative hypothesis that we are interested in. We might have a specific *a priori* prediction that if there were a departure from uniformity, then data points would be concentrated in a specific segment of the circle. For example, our prediction might be that most admissions to intensive care occur immediately after surgery, and the operating theatres are busiest between 9am and 5pm. If this were the case, we would be interested in testing against a specific alternative and this situation will be dealt with in Section 5.1.2. In Section 5.1.1 we consider contexts

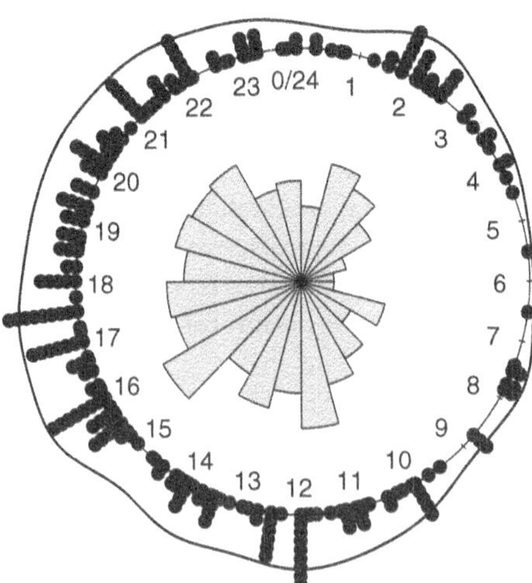

Figure 5.1 Circular plot of 254 patient arrival times at an intensive care unit together with a kernel density estimate and a rose diagram. The area of the sectors in the rose diagram represent the relative frequencies in the 24 class intervals

in which there is no *a priori* prediction about the position on the circle of any departure from uniformity. However, it is still important to consider whether you have an *a priori* expectation of the nature (rather than the position) of any departure.

5.1.1 Testing for Uniformity Against any Alternative

If you expect the departure from uniformity (should it occur) to involve only a single peak in the circular distribution, then the Rayleigh test should be used to specifically test for this type of departure from uniformity. For the intensive care admissions, it may be the case that staffing limitations are such that there has to be an eight-hour block of the day when the unit's ability to process new admissions is reduced, and you are interrogating the data to explore whether the timing of such a block of reduced availability should be decided on in part by making sure a full staff level is available at the time of day when demand is at a peak.

Alternatively, you may use so-called omnibus tests to test for a wider variety of types of departure from uniformity. Such tests will perform a little less well than the Rayleigh test in terms of their ability to detect unimodal deviations from uniformity, but have superior power with respect to more complex deviations. Our recommendation is that, unless you have a unimodal *a priori* prediction (as in the staffing limitation situation) or more complex departures would not be of interest, you should use omnibus tests.

In the first two testing scenarios considered below we assume the circular data are recorded on a continuous scale. In the third we discuss testing grouped circular data for uniformity.

The Rayleigh test for unimodal departures from uniformity

In Section 3.3.1 we introduced the mean resultant length, $\bar{R}$, as a measure of the concentration of data points around the circle. If $\bar{R}$ is greater than a threshold value, the data are interpreted as being too concentrated to be consistent with uniformity; and Rayleigh's test rejects the null hypothesis of uniformity. The Rayleigh test is known to be the most powerful invariant test for uniformity against von Mises alternatives (see Watson and Williams (1956)).

In **R**'s **circular** library, the Rayleigh test can be applied to the intensive care admissions data using the following commands:

```
rtest <- rayleigh.test(fisherB1c) ; rtest$statistic ; rtest$p
```

In response, **R** outputs a test statistic value of 0.3173 and a p-value for the test of 7.8305×10^{-12}. As the p-value is effectively zero, we emphatically reject the null hypothesis of uniformity and conclude that admissions to the intensive care unit are not spread evenly across the 24 hours of the day.

Omnibus tests

In contrast with testing against unimodal departures from uniformity, where the Rayleigh test is the one generally applied, there are various commonly used omnibus tests for uniformity. These include Ajne's A_n, Kuiper's V_n, Watson's U^2 and Rao's spacing test. These tests, and others, are described in detail in Batschelet (1981, Chapter 4), Upton and Fingleton (1989, Section 9.5) and Mardia and Jupp (1999, Section 6.3). The last three of the tests are available within **R**'s **circular** library. The limited comparisons which have been made of the various tests (Stephens, 1969; Bogdan *et al.*, 2002) suggest that there is no test that is superior to the others under all circumstances. Moreover, none of them have very high power, and several promising alternatives have recently been developed: see Bogdan *et al.* (2002) and Pycke (2010). Nevertheless, the commonly used omnibus tests for uniformity are all more powerful than Rayleigh's test when it comes to detecting multimodal departures from uniformity, and we encourage their use when there is no clear *a priori* reason to suspect (or to be predominantly interested in) unimodal departures rather than multimodal ones.

Using **R**'s **circular** library, the Kuiper, Watson and Rao spacing tests can be applied to the intensive care admissions data using the commands:

```
kuiper.test(fisherB1c) ; watson.test(fisherB1c) ; rao.spacing.test(fisherB1c)
```

In response, **R** quotes the p-values of the tests as being: < 0.01, < 0.01, and < 0.001, respectively. As with Rayleigh's test for unimodal departures, then, all three tests emphatically reject the null hypothesis of uniformity at, at least, the 1% significance level.

Omnibus tests for grouped data

The critical values of the tests for uniformity discussed so far were calculated under the assumption that the circular data being analysed are continuous. In our admission times example above, each time-point in the sample was actually recorded to the nearest minute. However, the 1440 minutes in a day are sufficiently numerous for the assumption of continuity to be a reasonable approximation. In practice, circular data are often aggregated into a smaller number of mutually exclusive categories. This is a common situation when data pertaining to the time of year are recorded as monthly totals. An example of such a data set was originally presented in Edwards (1961) and is reproduced in Table 5.1. It gives, for a particular UK city during a specific five-year period, the number of babies born per month with a focal condition.

Table 5.1 Monthly numbers of babies born, in a certain UK city during a specific five-year period, with a focal condition. The total number of such births is 176. The data first appeared in Edwards (1961).

Month	Births	Month	Births
January	10	July	7
February	19	August	10
March	18	September	13
April	15	October	23
May	11	November	15
June	13	December	22

When exploring whether there is evidence of any seasonal variation in such births, we will be interested in testing the null hypothesis of uniformity against all alternatives. Note that when analysing grouped data some authors simply use a chi-squared goodness-of-fit test. Whilst relatively easy to apply, the chi-squared test ignores the natural ordering of the categories (the order of the months in a year in the case of the focal condition births), and this can lead to reduced statistical power in comparison with other tests that do allow for such ordering. Steele and Chaseling (2006) provide a discussion of the issue.

There are numerous omnibus tests that have been proposed in the literature for testing grouped circular data for uniformity or, more generally, the goodness-of-fit of a (fully) specified distribution. Mardia and Jupp (1999, Chapter 6) discuss a version of Kuiper's test proposed by Freedman (1979), and three versions of Watson's test due to Freedman (1981), Choulakian *et al.* (1994) and Brown (1994), for use with grouped data. None of these tests have been programmed in **R**'s **circular** library, and critical values for their general use are not available. However, for goodness-of-fit testing of a specified distribution to grouped data, the significance of any of these tests can be established using the parametric bootstrap, with bootstrap samples drawn from the hypothesized distribution. Indeed, this is also true for any of the omnibus tests mentioned previously, designed originally to be used with continuous circular data. In order to illustrate the application of such an approach when testing grouped data for uniformity, we will consider the use of the adaptation of Watson's U^2 statistic due to Choulakian *et al.* (1994) and, because they are available within **R**'s **circular** library, the Kuiper and Watson omnibus tests designed originally to be employed with continuous data. Given its construction, which is based on the distances between the data points, it is not sensible to apply the Rao spacing test to grouped data.

Generalizing, for the moment, the situation involving the focal condition births, suppose we have a sample of circular data grouped into k mutually exclusive class intervals, $[\theta_{(0)}, \theta_{(1)}), [\theta_{(1)}, \theta_{(2)}), \ldots, [\theta_{(k-1)}, \theta_{(k)})$, where $\theta_{(0)} = \theta_{(k)} \pmod{2\pi}$, with n_j observations in the jth interval and thus a total of $n = n_1 + n_2 + \cdots + n_k$ observations. Suppose also that, for the distribution specified under the null hypothesis, the probability of a randomly chosen observation falling in the jth class interval is $p_j, j = 1, \ldots, k$. Then the test statistic for the adaptation of Watson's U^2 statistic due to Choulakian *et al.* (1994) is

$$U_G^2 = \frac{1}{n}\sum_{j=1}^{k} p_j(S_j - \bar{S})^2, \qquad (5.1)$$

where $S_j = \sum_{i=1}^{j}(n_i - np_i), j = 1, \ldots, k$, and $\bar{S} = \sum_{j=1}^{k} p_j S_j$.

In order to estimate the p-value of the test based on U_G^2 for data grouped into monthly totals (like the focal condition births) we use B bootstrap samples consisting of n days drawn uniformly and at random with replacement from the 365 days of the year and then grouped into the months of the year. The value of U_G^2 is calculated for the original data and for each bootstrap sample and the p-value of the test is then estimated by the proportion of the $(B + 1)$ values of U_G^2 that are at least as extreme as (here, greater than or equal to) the value of U_G^2 for the original data.

The function **UGsqMonTotalsBoot** below implements the bootstrap version of the test for a data set of monthly totals. The function makes no allowance for leap years but their effects should be minimal. Its initial lines set up various objects used subsequently in the calculation of (5.1), particularly probabilities and expected values under uniformity. Those objects are then called from the function **UGsq** written to compute the value of U_G^2. In the remainder of **UGsqMonTotalsBoot, UGsq** is used to compute the value of U_G^2 for the original data and B bootstrap samples, and the estimate of the p-value of the test is computed. Finally, the latter is returned together with all $(B + 1)$ values of U_G^2.

```
UGsqMonTotalsBoot<- function (montotals, B) {
nmon <- 12 ; mons <- seq(1:nmon)
daysmon <- c(31,28,31,30,31,30,31,31,30,31,30,31)
daysyear <- rep(mons,daysmon)
n <- sum(montotals) ; Pval <- daysmon/365 ; Eval <- Pval*n
UGsq <- function (mtots) {
Dval <- mtots-Eval ; Sval <- cumsum(Dval) ; Sbar <- sum(Pval*Sval)
tstat <- sum((Sval-Sbar)*(Sval-Sbar)*Pval)/n ; return(tstat)
}
tstat <- UGsq(montotals) ; nxtrm <- 1
for (b in 2:(B+1)) {
umontot <- 0 ; for (j in 1:nmon) { umontot[j] <- 0 }
udays <- sample(daysyear, size=n, replace = TRUE)
for (j in 1:n) { umontot[udays[j]] <- umontot[udays[j]]+1 }
tstat[b] <- UGsq(umontot)
if (tstat[b] >= tstat[1]) { nxtrm <- nxtrm+1}
}
pval <- nxtrm/(B+1) ; return(list(pval, tstat))
}
```

As explained in Section 1.10, the estimated p-value produced using this code will depend on the B randomly chosen bootstrap samples drawn when running the code. However, if B is large it should vary only slightly between runs.

We can run the bootstrap version of the test for the focal condition births data set and $B = 9999$ bootstrap samples, output the value of the test statistic for the original data and the estimated p-value, and produce a histogram of all 10, 000 computed values of U_G^2, using the following commands:

```
monbirths <- c(10,19,18,15,11,13,7,10,13,23,15,22)
B <- 9999 ; bootres <- UGsqMonTotalsBoot(monbirths, B)
pval <- bootres[[1]] ; UGsqval <- bootres[[2]] ; UGsqval[1] ; pval
hist(UGsqval, freq=FALSE, breaks=40, main=" ", xlab="UGsq value", ylab="Density")
```

In response, **R** outputs the value 0.2243 for U_G^2 and, when we ran this code, an estimated p-value of 0.0260. Using (1.1) with $\hat{p} = 0.0260$ and $N_R = B = 9999$ results in a 95% confidence interval for the true p-value of the test of $(0.0229, 0.0291)$. So it would appear that uniformity can safely be rejected at the 3% significance level and above. The histogram produced for the 10, 000 U_G^2-values is portrayed in Fig. 5.2.

As alternatives to the use of U_G^2, one can bootstrap the Kuiper and Watson omnibus tests for uniformity as implemented in **R**. This is done in the following function.

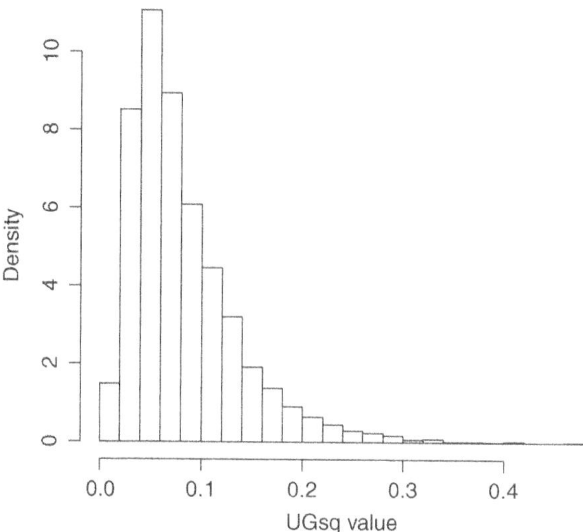

Figure 5.2 Histogram of the values of U_G^2 for the monthly focal condition births and $B = 9999$ bootstrap samples of the same size with days sampled uniformly across the year

```
KWMonTotalsBoot<- function (montotals, B) {
n <- sum(montotals)
daysmon <- c(31,28,31,30,31,30,31,31,30,31,30,31)
monuplim <- cumsum(daysmon) ; mtotuplim <- rep(monuplim, montotals)
cmtotuplim <- circular(mtotuplim*(2*pi/365))
kuistat <- kuiper.test(cmtotuplim)$statistic
watstat <- watson.test(cmtotuplim)$statistic
nxtrm <- 0 ; pval <- 0
for (k in 1:2) { nxtrm[k] <- 1 }
umonuplim <- rep(monuplim,daysmon)
cumonuplim <- circular(umonuplim*(2*pi/365))
for (b in 2:(B+1)) {
uuplim <- sample(cumonuplim, size=n, replace=TRUE)
kuistat[b] <- kuiper.test(uuplim)$statistic
watstat[b] <- watson.test(uuplim)$statistic
if (kuistat[b] >= kuistat[1]) { nxtrm[1] <- nxtrm[1]+1 }
if (watstat[b] >= watstat[1]) { nxtrm[2] <- nxtrm[2]+1 }
}
for (k in 1:2) { pval[k] <- nxtrm[k]/(B+1) }
return(pval)
}
```

The main difference between this function and **UGsqMonTotalsBoot** is that here upper limits of the class intervals, rather than days of the year, are selected uniformly to form the bootstrap samples. This is because the only unequivocal information about the data values in grouped data sets is the number of them that are less than or equal to the upper limits of the class intervals. When we ran **KWMonTotalsBoot**, the estimated p-values returned for the bootstrapped versions of the Kuiper and Watson tests were 0.0266 and 0.0245, respectively. Both p-values support the conclusions drawn from bootstrapping the U_G^2 test.

5.1.2 *Testing for Uniformity Against a Unimodal Alternative with a Specified Mean Direction*

Sometimes we will be interested in testing uniformity against not all potential unimodal alternatives, but rather one with a specified population mean direction, μ. Tests of such constrained alternative hypotheses are more powerful than those considered so far in this chapter for which the alternative is unconstrained.

For the data set of intensive care admissions considered previously in this chapter we might reasonably formulate such an alternative. We had speculated that most admissions result from planned major surgery, with most planned operations beginning at a time from 9am until 5pm. If these operations last on average for two hours, then we would expect an increase in admissions between 11am and 7pm, with the mean for these extra admissions being in the middle of that period, at 3pm. Hence, we might have reasonably decided to test the null hypothesis of homogenous admissions over the 24-hour day against a unimodal alternative with a mean direction corresponding to 15:00 hours. This can be tested using a modified version of the Rayleigh test introduced in Section 5.1.1.

With the basic version of Rayleigh's test, if the sample mean resultant length, $\bar{R}$, is greater than a threshold value, the data are interpreted as being too concentrated to be consistent with uniformity, and the null hypothesis of uniformity is rejected. For the modified version of the Rayleigh test, involving the difference between the specified population mean direction, μ, and the sample mean direction, $\bar{\theta}$, the test statistic becomes $\bar{R}\cos(\bar{\theta} - \mu)$.

In **R**'s **circular** library, we can apply the modified Rayleigh test for the intensive care admission times and a specified mean direction corresponding to 15:00 hours by running:

```
mu <- circular(15, units="hours", template="clock24")
rayleigh.test(fisherB1c, mu)
```

In response, **R** returns a test statistic value of 0.2635 and a p-value rounded to 0. Thus, we emphatically reject the null hypothesis of uniformity in favour of an underlying distribution with a mean direction corresponding to 15:00 hours. However, rejection of the null hypothesis of uniformity should not be interpreted as evidence that the population mean direction corresponds necessarily to 3pm. This caution is discussed in full by Aneshansley and Larkin (1981). Appropriate methods for testing for a specific mean direction are considered in Section 5.3.3.

5.2 Testing for Reflective Symmetry

As we saw in Section 4.2.4, there are two types of 'symmetry' on the circle: reflective symmetry and ℓ-fold symmetry. As a consequence, there are four testing scenarios involving symmetry which might be of interest when analysing circular data. The first, that of testing the null hypothesis of ℓ-fold symmetry, was considered by Jupp and Spurr (1983). They introduced rank-based test procedures to tackle this inferential problem. The second, involving testing for reflective symmetry about a specified axis, was considered by Schach (1969). He derived results for locally most powerful rank tests against rotation alternatives.

The third scenario, involving testing for reflective symmetry about a known median, was addressed by Pewsey (2004*b*). He proposed a hybrid testing strategy incorporating a new test and the circular analogue of the modified runs test of Modarres and Gastwirth (1996). In the remainder of this section we consider in detail the fourth scenario, that of testing the null hypothesis of an underlying distribution which is reflectively symmetric about an unknown central direction. This null hypothesis is arguably the second most important dividing hypothesis in circular data analysis, after that of isotropy. Its acceptance or rejection will greatly determine the forms of distributions we subsequently investigate to model our data.

5.2.1 *Large-sample Test for Reflective Symmetry*

The asymptotic result presented in Section 4.2.5 provides the theoretical basis of the test for reflective symmetry about an unknown central direction of Pewsey (2002*b*). According to that result, for an underlying circular distribution with mean resultant length $\rho \in (0, 1)$, the distribution of $\bar{b}_2$, the second sample sine moment about the sample mean direction $\bar{\theta}$, is asymptotically normally distributed with, to $O(n^{-3/2})$, mean

$$\bar{\beta}_2 + \frac{1}{n\rho}\left(-\bar{\beta}_3 - \frac{\bar{\beta}_2}{\rho} + \frac{2\bar{\alpha}_2\bar{\beta}_2}{\rho^3}\right), \tag{5.2}$$

and variance

$$\frac{1}{n}\left[\frac{1-\bar{\alpha}_4}{2} - 2\bar{\alpha}_2 - \bar{\beta}_2^2 + \frac{2\bar{\alpha}_2}{\rho}\left\{\bar{\alpha}_3 + \frac{\bar{\alpha}_2(1-\bar{\alpha}_2)}{\rho}\right\}\right]. \tag{5.3}$$

For a reflectively symmetric distribution with $\rho \in (0, 1)$, the mean direction, μ, exists and the sine moments about μ of any order, and hence (5.2), are 0. These results led Pewsey (2002*b*) to suggest an asymptotically distribution-free test of the null hypothesis of circular reflective symmetry about an unknown central direction based on the studentized statistic

$$z = \frac{\bar{b}_2}{\sqrt{\widehat{\text{var}}(\bar{b}_2)}}, \tag{5.4}$$

where $\widehat{\text{var}}(\bar{b}_2)$ is a plug-in estimate of (5.3) for an assumed reflectively symmetric population given by

$$\widehat{\text{var}}(\bar{b}_2) = \frac{1}{n}\left[\frac{1-\bar{a}_4}{2} - 2\bar{a}_2 + \frac{2\bar{a}_2}{\bar{R}}\left\{\bar{a}_3 + \frac{\bar{a}_2(1-\bar{a}_2)}{\bar{R}}\right\}\right].$$

Large absolute values of (5.4) compared with the quantiles of the standard normal distribution lead to the rejection of reflective symmetry in favour of some skewed alternative. Simulation results reported in Pewsey (2002*b*) suggest this asymptotic theory based test can be used for sample sizes of 50 or more. For smaller sample sizes, the bootstrap version of the test described in Section 5.2.2 is recommended. Both versions of the test assume that the circular data being analysed are represented as angles measured in radians.

Henceforth, the objects **circdat, cdat** and **origdat** are assumed to be circular data objects containing values measured in radians in $[0, 2\pi)$. The following function calculates the absolute value of the test statistic (5.4) for the object **circdat**:

```
RSTestStat <- function(circdat) {
n <- length(circdat) ; Rbar <- rho.circular(circdat)
t2bar <- trigonometric.moment(circdat, p=2, center=TRUE)
t3bar <- trigonometric.moment(circdat, p=3, center=TRUE)
t4bar <- trigonometric.moment(circdat, p=4, center=TRUE)
bbar2 <- t2bar$sin ; abar2 <- t2bar$cos
abar3 <- t3bar$cos ; abar4 <- t4bar$cos
var <- ((1-abar4)/2-(2*abar2)+(2*abar2/Rbar)*(abar3+(abar2*(1-abar2)/Rbar)))/n
absz <- abs(bbar2/sqrt(var)) ; return(absz)
}
```

To illustrate the use of this new function when performing the large-sample test, we run it with the data set introduced in Section 2.2 consisting of 310 wind directions recorded in the Italian Alps. This can be achieved using the following commands:

```
cdat <- circular(wind) ; absz <- RSTestStat(cdat)
pval <- 2*pnorm(absz, mean=0, sd=1, lower=FALSE) ; pval
```

The p-value returned by **R** is 4.7720×10^{-8}, and hence the null hypothesis of circular reflective symmetry is emphatically rejected. In contrast, for the intensive care unit admissions data introduced in Section 5.1.1, the code:

```
cdat <- circular(fisherB1*2*pi/24) ; absz <- RSTestStat(cdat)
pval <- 2*pnorm(absz, mean=0, sd=1, lower=FALSE) ; pval
```

returns a p-value of 0.2090. Thus, for these data, it appears reasonable to assume that the underlying distribution is reflectively symmetric.

5.2.2 *Bootstrap Test for Reflective Symmetry*

When applying the bootstrap version of the test we make use of the following symmetrization device due to Efron (1979). For the original sample of size n, $\theta_1, \ldots, \theta_n$, we first calculate the mean direction, $\bar{\theta}$. As the deviation of the angle θ_j from $\bar{\theta}$ is $\theta_j - \bar{\theta}$, the reflection of θ_j about $\bar{\theta}$ is given by $\bar{\theta} - (\theta_j - \bar{\theta}) = 2\bar{\theta} - \theta_j$. A symmetrized version of the original sample, of size $2n$, is then given by $\{\theta_1, \ldots, \theta_n, 2\bar{\theta} - \theta_1, \ldots, 2\bar{\theta} - \theta_n\}$. To test for reflective symmetry of the underlying distribution from which the original sample was drawn, we draw B bootstrap samples of size n from the symmetrized sample and compute the absolute value of (5.4) for each one. The p-value of the test is estimated by the proportion of the $(B + 1)$ values of (5.4) that are greater than or equal to the absolute value of (5.4) for the original sample. This testing approach is programmed in the function **RSTestBoot** which makes use of the previously defined function **RSTestStat**.

```
RSTestBoot<- function(origdat, B) {
n <- length(origdat) ; absz <- RSTestStat(origdat)
tbar <- mean(origdat) ; refcdat <- 2*tbar-origdat
symmcdat <- c(origdat, refcdat) ; nxtrm <- 1
for (b in 2:(B+1)) {
bootsymmdat <- sample(symmcdat, size=n, replace=TRUE)
```

```
absz[b] <– RSTestStat(bootsymmdat)
if (absz[b] >= absz[1]) {nxtrm <– nxtrm+1}
}
pval <– nxtrm/(B+1) ; return(pval)
}
```

To illustrate the use of **RSTestBoot**, we run it with the **fisherB6$set1** data object available within the **circular** package. This data set consists of $n = 40$ cross-bed azimuths of palaeocurrents measured, in degrees clockwise from north, in the Belford Anticline, New South Wales, Australia. A circular data plot of the cross-bed azimuths converted to radians and represented as angles measured counterclockwise from 0 is portrayed in Fig. 5.3. With a p-value quoted as being less than 0.01, Kuiper's test rejects isotropy for these data. We can apply the bootstrap implementation of the test for reflective symmetry using the commands:

```
cdat <– circular(fisherB6$set1*2*pi/360) ; B <– 9999
pval <– RSTestBoot(cdat, B) ; pval
```

When we ran this code, the estimated p-value returned by **R** was 0.5391. Using (1.1), a 95% confidence interval for the true p-value is $(0.5293, 0.5489)$, so we have no reason to reject the null hypothesis of reflective symmetry for the underlying distribution from which the data were drawn.

We have presented two approaches to testing for reflective symmetry: one based on asymptotic normal theory and the other using bootstrap samples drawn from a symmetrized version of the data under investigation. As explained, the large-sample version of the test should only be used when the sample size is 50 or more. For such sample sizes its ability to hold the nominal level, and its power, are competitive with those of the bootstrap version of the test. It is also considerably faster to execute as it does not require any resampling.

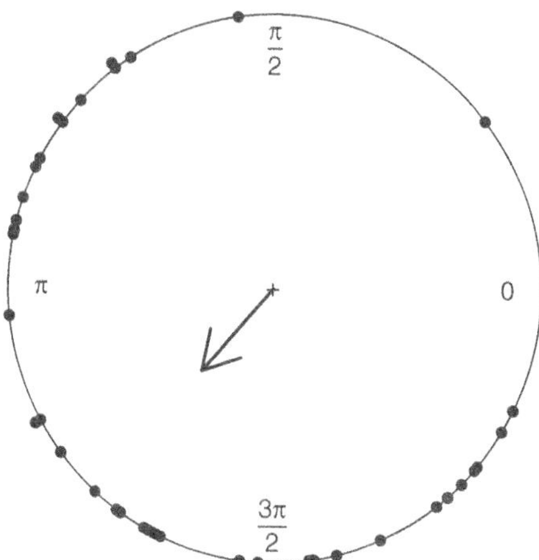

Figure 5.3 Circular data plot of 40 palaeocurrent cross-bed azimuths represented in radians measured counterclockwise from 0. The arrow represents the mean resultant vector

Both versions of the test based on (5.4) assume that, for the underlying distribution, $\rho \in (0, 1)$. When $\rho = 1$ the underlying distribution is a point distribution and hence there is no need to test for reflective symmetry. If $\rho = 0$ then the underlying distribution is circular uniform or a multimodal distribution which is either cyclically symmetric, or has more than one axis of reflective symmetry, or both. As argued at the beginning of this section, we will only be interested in testing for circular reflective symmetry if uniformity has previously been rejected. If it is thought possible that $\rho = 0$ for some underlying multimodal distribution, and the number of modes, m, of the distribution can be established beforehand, then the appropriate version of the test can be used after first applying the device of m-fold wrapping of the circle onto itself (see, for example, Mardia and Jupp (1999, page 53)). This device involves multiplying each data value by m, by so doing producing a sample from a distribution with a uniquely defined mean direction. This procedure can, in fact, be used with data from any form of multimodal distribution, and so if there is any doubt as to whether $\rho \neq 0$ it is advisable to apply the appropriate version of the test to both the original data and to the wrapped data and compare the resulting p-values.

5.3 Inference for Key Circular Summaries

So far we have seen how, in the initial inferential stage of circular statistical analysis, we can perform formal tests designed to explore the fundamental dividing hypotheses of isotropy and reflective symmetry. Next, prior to fitting a model to our data, we will generally be interested in performing distribution-free inference for population summaries such as the mean direction, μ, mean resultant length, ρ, second central sine moment, $\bar{\beta}_2$, and second central cosine moment, $\bar{\alpha}_2$. As was explained in Section 4.2.3, these are the fundamental circular measures of location, concentration, skewness and kurtosis. By analogy with Section 5.2, the asymptotic results for the asymptotic distribution of $\bar{\zeta} = (\bar{\theta}, \bar{R}, \bar{b}_2, \bar{a}_2)^T$ of Pewsey (2004*a*), summarized in Section 4.2.5, provide a basis with which to carry out inference for large-sized samples, whilst the bootstrap is available when working with samples containing fewer observations. We begin by considering bias-corrected point estimation and confidence interval construction for the four population summaries before proceeding to hypothesis testing for specific values of the mean direction, μ. To simplify the presentation, the methods and **R** code presented assume that the circular data being analysed are represented as angles lying in $[0, 2\pi)$ measured in radians counterclockwise from zero. If the original data were not of this type they should be converted to such angles prior to applying the methods described here. As we will illustrate in the examples, any results involving μ or $\bar{\beta}_2$ should be 'back-transformed' when relating them to the underlying distribution for the original data.

5.3.1 *Bias-corrected Point Estimation*

The plug-in estimates of μ, ρ, $\bar{\beta}_2$ and $\bar{\alpha}_2$ are simply their sample analogues $\bar{\theta}$, $\bar{R}$, $\bar{b}_2$ and $\bar{a}_2$. However, the large-sample results in (4.29) and (4.30) indicate that all four estimates are generally biased. Substituting unknown population summaries by their sample analogues as appropriate, a bias-corrected point estimate for μ is

$$\hat{\mu}_{BC} = \bar{\theta} - \widehat{\text{bias}}(\bar{\theta}), \quad \text{where} \quad \widehat{\text{bias}}(\bar{\theta}) = -\frac{\bar{b}_2}{2n\bar{R}^2}. \tag{5.5}$$

Similarly, bias-corrected point estimates for ρ, $\bar{\beta}_2$ and $\bar{\alpha}_2$ are given by:

$$\hat{\rho}_{BC} = \bar{R} - \widehat{\text{bias}}(\bar{R}), \quad \text{where} \quad \widehat{\text{bias}}(\bar{R}) = \frac{1-\bar{a}_2}{4n\bar{R}},$$

$$\widehat{\bar{\beta}}_{2BC} = \bar{b}_2 - \widehat{\text{bias}}(\bar{b}_2), \quad \text{where} \quad \widehat{\text{bias}}(\bar{b}_2) = \frac{1}{n\bar{R}}\left(-\bar{b}_3 - \frac{\bar{b}_2}{\bar{R}} + \frac{2\bar{a}_2\bar{b}_2}{\bar{R}^3}\right),$$

and

$$\widehat{\bar{\alpha}}_{2BC} = \bar{a}_2 - \widehat{\text{bias}}(\bar{a}_2),$$

where

$$\widehat{\text{bias}}(\bar{a}_2) = \frac{1}{n}\left\{1 - \frac{\bar{a}_3}{\bar{R}} - \frac{\bar{a}_2(1-\bar{a}_2) + \bar{b}_2^2}{\bar{R}^2}\right\}, \tag{5.6}$$

respectively. Clearly, if the underlying distribution is assumed to be reflectively symmetric then inference for $\bar{\beta}_2$ will be of no interest. Testing for circular reflective symmetry was considered in Section 5.2. Under the assumption of a reflectively symmetric underlying distribution, the second central sine moment appearing in Equations (5.5) and (5.6) should be set equal to 0. **R** code to calculate these bias-corrected point estimates is included in the next subsection.

5.3.2 *Bias-corrected Confidence Intervals*

Building on the content of Section 5.3.1, and again using the results in (4.29) and (4.30) for the asymptotic distribution of $\bar{\zeta} = (\bar{\theta}, \bar{R}, \bar{b}_2, \bar{a}_2)^T$ with any unknown population summaries substituted by their sample analogues as appropriate, a nominally $100(1-\alpha)\%$ bias-corrected confidence interval for μ is given by

$$\hat{\mu}_{BC} \pm z_{(1-\alpha/2)}\{\widehat{\text{var}}(\bar{\theta})\}^{1/2}, \tag{5.7}$$

where $z_{(1-\alpha/2)}$ denotes the $(1-\alpha/2)$ quantile of the standard normal distribution, and

$$\widehat{\text{var}}(\bar{\theta}) = \frac{1-\bar{a}_2}{2n\bar{R}^2}. \tag{5.8}$$

Similarly, nominally $100(1-\alpha)\%$ bias-corrected confidence intervals for ρ, $\bar{\beta}_2$ and $\bar{\alpha}_2$ are given by:

$$\hat{\rho}_{BC} \pm z_{(1-\alpha/2)}\{\widehat{\text{var}}(\bar{R})\}^{1/2}, \quad \text{where} \quad \widehat{\text{var}}(\bar{R}) = \frac{1-2\bar{R}^2+\bar{a}_2}{2n},$$

$$\widehat{\bar{\beta}}_{2BC} \pm z_{(1-\alpha/2)}\{\widehat{\text{var}}(\bar{b}_2)\}^{1/2},$$

where

$$\widehat{\text{var}}(\bar{b}_2) = \frac{1}{n}\left[\frac{1-\bar{a}_4}{2} - 2\bar{a}_2 - \bar{b}_2^2 + \frac{2\bar{a}_2}{\bar{R}}\left\{\bar{a}_3 + \frac{\bar{a}_2(1-\bar{a}_2)}{\bar{R}}\right\}\right],$$

and

$$\widehat{\bar{\alpha}}_{2BC} \pm z_{(1-\alpha/2)}\{\widehat{\text{var}}(\bar{a}_2)\}^{1/2},$$

where

$$\widehat{\text{var}}(\bar{a}_2) = \frac{1}{n}\left[\frac{1-2\bar{a}_2^2+\bar{a}_4}{2} + \frac{2\bar{b}_2}{\bar{R}}\left\{\bar{b}_3 + \frac{\bar{b}_2(1-\bar{a}_2)}{\bar{R}}\right\}\right], \quad (5.9)$$

respectively. Again, if the underlying distribution is assumed to be reflectively symmetric then inference for $\bar{\beta}_2$ will be of no interest. Under the assumption of a reflectively symmetric underlying distribution, the central sine moments in Equation (5.9) should be set equal to 0.

The function **ConfIntLS** below computes bias-corrected point estimates and nominally **conflevel**% confidence intervals for μ, ρ and $\bar{\alpha}_2$, and for $\bar{\beta}_2$ when the underlying distribution is assumed not to be reflectively symmetric. Its **indsym** argument is used to indicate whether the data in the circular data object **circdat** are assumed to have been drawn from an underlying distribution which is reflectively symmetric (**indsym = 1**) or asymmetric (**indsym = 0**).

```
ConfIntLS <- function(circdat, indsym, conflevel) {
n <- length(circdat) ; tbar <- mean(circdat) ; Rbar <- rho.circular(circdat)
t2bar <- trigonometric.moment(circdat, p=2, center=TRUE)
t3bar <- trigonometric.moment(circdat, p=3, center=TRUE)
t4bar <- trigonometric.moment(circdat, p=4, center=TRUE)
abar2 <- t2bar$cos ; abar3 <- t3bar$cos ; abar4 <- t4bar$cos
bbar2 <- t2bar$sin ; bbar3 <- t3bar$sin
Rbar2 <- Rbar*Rbar ; Rbar4 <- Rbar2*Rbar2
alpha <- (100-conflevel)/100 ; qval <- qnorm(1-alpha/2)
rhobc <- Rbar - ((1-abar2)/(4*n*Rbar))
rbarstderr <- sqrt((1-2*Rbar2+abar2)/(2*n))
rhoup <- rhobc+qval*rbarstderr ; rholo <- rhobc-qval*rbarstderr
rhores <- c(rhobc, rholo, rhoup)
if (indsym == 1) { bbar2 <- 0 ; bbar3 <- 0 } else
if (indsym == 0) {
betab2bc <- bbar2 + ((bbar3/Rbar)+(bbar2/Rbar2)-(2*abar2*bbar2/Rbar4))/n
```

```
b2bstderr <- sqrt((((1-abar4)/2)-(2*abar2)-(bbar2*bbar2)+(2*abar2/Rbar)*
(abar3+(abar2*(1-abar2)/Rbar)))/n)
betab2up <- betab2bc+qval*b2bstderr
betab2lo <- betab2bc-qval*b2bstderr
betab2res <- c(betab2bc, betab2lo, betab2up)
}
div <- 2*n*Rbar2
mubc <- tbar+(bbar2/div) ; tbarstderr <- sqrt((1-abar2)/div)
muup <- mubc+qval*tbarstderr ; mulo <- mubc-qval*tbarstderr
mures <- c(mubc, mulo, muup)
alphab2bc <- abar2-(1-(abar3/Rbar)-((abar2*(1-abar2)+bbar2*bbar2)/Rbar2))/n
a2bstderr <- sqrt((((1+abar4)/2)-(abar2*abar2)+(2*bbar2/Rbar)*(bbar3+(bbar2*(1-abar2)/Rbar)))/n)
alphab2up <- alphab2bc+qval*a2bstderr
alphab2lo <- alphab2bc-qval*a2bstderr
alphab2res <- c(alphab2bc, alphab2lo, alphab2up)
if (indsym == 0) { return(list(mures, rhores, betab2res, alphab2res)) } else
if (indsym == 1) { return(list(mures, rhores, alphab2res)) }
}
```

To illustrate the use of **ConfIntLS**, we apply it to the Italian Alps wind directions and intensive care arrival times data sets. In both cases, the sample size is considerably larger than 50. Remember, in Section 5.2 the large-sample version of the test for reflective symmetry rejected reflective symmetry for the underlying distribution of the wind directions but not for the arrival times. For the wind directions treated as being measured counterclockwise from 0 (north), the **R** code:

```
cdat <- circular(wind) ; sym <- 0 ; clev <- 95
LSCIOut <- ConfIntLS(cdat, sym, clev) ; LSCIOut
```

returns bias-corrected point estimates for μ, ρ, $\bar{\beta}_2$ and $\bar{\alpha}_2$ of, to two decimal places, 0.29 (radians measured clockwise from north for the original wind directions), 0.66, −0.20 and 0.43, respectively. The nominally 95% confidence intervals returned for the same population measures are $(0.20, 0.38)$ radians, $(0.60, 0.71)$, $(-0.27, -0.13)$, and $(0.34, 0.52)$, respectively. The point estimate and interval for $\bar{\beta}_2$ transform to values of 0.20 and $(0.13, 0.27)$, respectively, for the original wind directions. The intervals for ρ and $\bar{\beta}_2$ do not contain the value 0, supporting the earlier rejection of both isotropy and reflective symmetry for the underlying distribution. The interval for μ corresponds to a relatively short arc subtending an angle of approximately 10.4°. Thus the central location of the underlying distribution is estimated fairly precisely. Finally, the interval for $\bar{\alpha}_2$ suggests that the underlying distribution is considerably more peaked than a wrapped normal distribution with a ρ-value estimated to be 0.66, for which, using Equations (4.19) and (4.49), a point estimate of $\bar{\alpha}_2$ would be $(0.66)^4 = 0.18$.

For the intensive care admissions data, for which the assumption of a reflectively symmetric underlying distribution appears reasonable, running the code:

```
cdat <- circular(fisherB1*2*pi/24)
sym <- 1 ; clev <- 95
LSCIOut <- ConfIntLS(cdat, sym, clev) ; LSCIOut
```

produces point estimates for μ, ρ and $\bar{\alpha}_2$ of 4.49 (radians clockwise from midnight for the original arrivals data, i.e. around 5:09 pm), 0.31 and −0.05, respectively. The nominally 95%

confidence intervals outputted are $(4.21, 4.77)$ radians, $(0.24, 0.39)$ and $(-0.13, 0.04)$, respectively. The interval for ρ does not contain 0, supporting our earlier rejection of isotropy for the underlying distribution. The interval for μ is relatively wide, corresponding to a time period of approximately 2 hours and 8 minutes. According to it, the mean admission time could be anywhere between 4.05pm and 6.13pm. In Section 5.1.2 we rejected uniformity for these data in favour of a unimodal distribution with a mean direction corresponding to 3pm. However, we also cautioned against interpreting that rejection as evidence that the population mean direction corresponds necessarily to 3pm. Here we see that the confidence interval for μ does not include 3pm, and hence we reject it, at the 5% significance level, as corresponding to the mean direction of the underlying distribution. Hence, our earlier cautionary remarks regarding the interpretation of the results from the test of uniformity were justified. In fact, as we shall see in Section 5.3.3, the null hypothesis of a mean direction corresponding to 3pm can be rejected at far below the 5% level. Finally, the interval for $\bar{\alpha}_2$ suggests the underlying distribution is probably more flat-topped than a wrapped normal distribution with an estimated ρ-value of 0.31, for which a point estimate of $\bar{\alpha}_2$ would be $(0.31)^4 = 0.01$.

The bootstrap counterparts of the large-sample confidence intervals described above can be computed using the techniques introduced previously in this chapter. The bootstrap samples are drawn from the original data set if reflective symmetry was rejected by the test described in Section 5.2. If reflective symmetry was not rejected then they are drawn from the symmetrized sample instead. For the original sample and each of B bootstrap samples, the bias-corrected estimates of the population measures are computed using the function **BiasCEsts**, and the estimates of each measure ordered. As can be seen in the function **ConfIntBoot** written to perform the necessary bootstrap computations, the calculation of the interval for μ requires the most care. In particular, note that the distances calculated between the bias-corrected estimate of μ for the original sample and those for the bootstrap samples are as defined in Equation (3.19). If the underlying distribution is assumed to be symmetric, a symmetric confidence interval for μ is computed. Otherwise, $(1-\alpha)100\%$ confidence intervals are estimated using the $\alpha/2$ and $(1-\alpha/2)$ quantiles of the bootstrap sampling distribution of each bias-corrected estimate. As previously, no inference is performed for $\bar{\beta}_2$ if the underlying distribution is assumed to be reflectively symmetric.

```
BiasCEsts <- function(circdat, indsym, n) {
t10bar <- trigonometric.moment(circdat, p=1, center=FALSE)
tbar <- atan2(t10bar$sin, t10bar$cos) ; if (tbar < 0) {tbar <- tbar+2*pi}
Rbar <- rho.circular(circdat)
t2bar <- trigonometric.moment(circdat, p=2, center=TRUE)
t3bar <- trigonometric.moment(circdat, p=3, center=TRUE)
abar2 <- t2bar$cos ; abar3 <- t3bar$cos
bbar2 <- t2bar$sin ; bbar3 <- t3bar$sin
Rbar2 <- Rbar*Rbar ; Rbar4 <- Rbar2*Rbar2
rhobc <- Rbar - ((1-abar2)/(4*n*Rbar))
if (indsym == 1) {bbar2 <- 0 ; bbar3 <- 0 ; betab2bc <- 0} else
if (indsym == 0) {
betab2bc <- bbar2 + ((bbar3/Rbar)+(bbar2/Rbar2)-(2*abar2*bbar2/Rbar4))/n
}
div <- 2*n*Rbar2 ; mubc <- tbar+(bbar2/div)
```

```
if (mubc > 2*pi) {mubc <- mubc-2*pi} else
if (mubc < 0) {mubc <- mubc+2*pi}
alphab2bc <- abar2- (1-(abar3/Rbar)-((abar2*(1-abar2)+bbar2*bbar2)/Rbar2))/n
return(list(mubc, rhobc, betab2bc, alphab2bc))
}

ConfIntBoot <- function(origdat, indsym, conflevel, B) {
alpha <- (100-conflevel)/100 ; n <- length(origdat)
ests <- BiasCEsts(origdat, indsym, n)
muest <- ests[[1]] ; rhoest <- ests[[2]]
betab2est <- ests[[3]] ; alphab2est <- ests[[4]]
if (indsym == 1) {
refdat <- 2*muest-origdat ; sampledat <- c(origdat, refdat) } else
if (indsym == 0) { sampledat <- origdat }
for (b in 2:(B+1)) {
bootdat <- sample(sampledat, size=n, replace=TRUE)
ests <- BiasCEsts(bootdat, indsym, n)
muest[b] <- ests[[1]] ; rhoest[b] <- ests[[2]]
betab2est[b] <- ests[[3]] ; alphab2est[b] <- ests[[4]]
}
dist <- 0
if (indsym == 1) {
dist <- pi-abs(pi-abs(muest-muest[1])) ; sdist <- sort(dist)
mulo <- muest[1]-sdist[(B+1)*(1-alpha)]
muup <- muest[1]+sdist[(B+1)*(1-alpha)]
} else
if (indsym == 0) {
if (muest[1] < pi) {
ref <- muest[1]+pi
for (b in 1:(B+1)) {
dist[b] <- -(pi-abs(pi-abs(muest[b]-muest[1])))
if (muest[b] > muest[1]) {
if (muest[b] < ref) { dist[b] <- -dist[b] }
}
}
} else
if (muest[1] >= pi) {
ref <- muest[1]-pi
for (b in 1:(B+1)) {
dist[b] <- pi-abs(pi-abs(muest[b]-muest[1]))
if (muest[b] > ref) {
if (muest[b] < muest[1]) { dist[b] <- -dist[j] }
}
}
}
sdist <- sort(dist) ; mulo <- muest[1]+sdist[(B+1)*(alpha/2)]
muup <- muest[1]+sdist[(B+1)*(1-alpha/2)]
sbetab2est <- sort(betab2est)
betab2lo <- sbetab2est[(B+1)*(alpha/2)]
betab2up <- sbetab2est[(B+1)*(1-alpha/2)]
betab2res <- c(betab2est[1], betab2lo, betab2up)
}
mures <- c(muest[1], mulo, muup) ; srhoest <- sort(rhoest)
rholo <- srhoest[(B+1)*(alpha/2)] ; rhoup <- srhoest[(B+1)*(1-alpha/2)]
salphab2est <- sort(alphab2est)
alphab2lo <- salphab2est[(B+1)*(alpha/2)]
alphab2up <- salphab2est[(B+1)*(1-alpha/2)]
rhores <- c(rhoest[1], rholo, rhoup)
```

```
alphab2res <- c(alphab2est[1], alphab2lo, alphab2up)
if (indsym == 0) { return(list(mures, rhores, betab2res, alphab2res)) } else
if (indsym == 1) { return(list(mures, rhores, alphab2res)) }
}
```

The bootstrap version of the test in Section 5.2 did not reject the null hypothesis of an underlying reflectively symmetric distribution for the palaeocurrent cross-bed azimuth data. We now make use of the functions **BiasCEsts** and **ConfIntBoot** to obtain point estimates and nominally 95% confidence intervals for μ, ρ and $\bar{\alpha}_2$. The **R** code required to produce them is:

```
cdat <- circular(fisherB6$set1*2*pi/360)
sym <- 1 ; clev <- 95 ; B <- 9999
BCIOut <- ConfIntBoot(cdat, sym, clev, B) ; BCIOut
```

The point estimates returned for μ, ρ and $\bar{\alpha}_2$ are 3.98 radians (or approximately 228° clockwise from north for the original data), 0.39 and −0.21, respectively. When we ran the above code, the nominally 95% confidence intervals returned were $(3.41, 4.55)$ radians, $(0.23, 0.56)$ and $(-0.50, 0.13)$, respectively. The interval for ρ does not contain 0, so supporting the earlier rejection of isotropy for the underlying distribution. The relatively wide interval for μ, corresponding to an arc subtending approximately 65°, reflects the relatively low concentration of the data. It suggests that the mean azimuth could be anywhere between 195° and 261° clockwise from north. Finally, the interval for $\bar{\alpha}_2$ indicates that the underlying distribution could be as peaked as a wrapped normal distribution with an estimated ρ-value of 0.39, for which a point estimate of $\bar{\alpha}_2$ would be $(0.39)^4 = 0.02$. Nevertheless, a more flat-topped distribution appears the more likely. In Chapter 6 we will identify a model which fits this data set well.

5.3.3 *Testing for a Specified Mean Direction*

Using the results in Section 4.2.5 it is simple to define large-sample tests for specified values of any of the four population measures μ, ρ, $\bar{\beta}_2$ and $\bar{\alpha}_2$. Their bootstrap counterparts follow straightforwardly using the resampling techniques introduced in this chapter. In fact, we have already seen tests for specific values of two of the measures: the Rayleigh test of Section 5.1.1 tests the null hypothesis that $\rho = 0$, and Pewsey's test of reflective symmetry, considered in Section 5.2, tests the null hypothesis that $\bar{\beta}_2 = 0$. Whilst it is possible to generalize those two tests to test for other specified values, it is doubtful whether in practice one would actually be interested in testing for values other than 0.

It should also be remembered that in Section 5.3.2 we saw how to calculate individual confidence intervals for all four measures. Having calculated a $100(1-\alpha)\%$ confidence interval for a population measure, we can perform a test of the null hypothesis that the measure in question takes a specified value, at the $100\alpha\%$ significance level, by simply inspecting the confidence interval to see whether it contains the specified value. If the confidence interval contains the specified value of the measure then the data support it as a *potential* value for the underlying distribution. However, one cannot conclude that the specified value is necessarily the true one because all the values included in the confidence interval are potential values of the population measure. On the other hand, if the confidence interval does not contain the specified value then the null hypothesis is rejected at the $100\alpha\%$ significance level. For instance, at the end of Section 5.3.1, the nominally 95% confidence interval for $\bar{\alpha}_2$ of the distribution underlying the cross-bed azimuths was $(-0.50, 0.13)$. As this interval contains the value 0, we would not reject the null hypothesis that $\bar{\alpha}_2 = 0$ at (nominally) the 5% significance level. However, we would if the null hypothesis were that $\bar{\alpha}_2 = 0.2$. If we wanted to perform a test at some significance level other than 5%, say $100\alpha_0\%$, then we would have to compute a confidence interval at the $100(1-\alpha_0)\%$ confidence level. More generally, by iteratively calculating confidence intervals with different confidence levels, it would be possible to obtain an estimate of the p-value of a test for a specified value of a population measure. However, such an approach is far more time-consuming than calculating the p-value directly, obtained by comparing the observed value of a test statistic with its sampling distribution under the null hypothesis.

In the remainder of this subsection we consider testing for a specified value, μ_0, of the mean direction of an underlying circular distribution. This is the testing scenario of most relevance because in applications the mean direction will generally be the population measure of main interest. Throughout we assume that, like the data themselves, any directions are represented as angles lying in $[0, 2\pi)$ measured in radians counterclockwise from zero.

Using the results in Section 4.2.5, together with Equations (5.5), (5.7) and (5.8), a test of the null hypothesis that the mean direction of a circular distribution with $\rho \in (0, 1)$ is μ_0 can be based on the test statistic

$$z = \frac{\pi - |\pi - |\hat{\mu}_{BC} - \mu_0||}{\{\widehat{\mathrm{var}}(\bar{\theta})\}^{1/2}}, \tag{5.10}$$

where the numerator is the angular distance (3.19) between $\hat{\mu}_{BC}$ and μ_0. With this definition, the value taken by (5.10) is always non-negative. Large values of (5.10) compared with the quantiles of the standard normal distribution lead to the rejection of the null hypothesis in favour of an underlying distribution with a mean direction that is different from μ_0.

For the circular data object **circdat**, the **R** function **SpecMeanTestRes** below outputs the value of the test statistic (5.10) for an hypothesized population mean direction **mu0**. Its **indsym** argument is used to indicate whether the data are assumed to have been drawn from an underlying distribution which is reflectively symmetric (**indsym = 1**) or skewed (**indsym = 0**).

```
SpecMeanTestRes <- function(circdat, indsym, mu0) {
n <- length(circdat)
t10bar <- trigonometric.moment(circdat, p=1, center=FALSE)
tbar <- atan2(t10bar$sin, t10bar$cos)
if (tbar < 0) {tbar <- tbar+2*pi}
Rbar <- rho.circular(circdat) ; Rbar2 <- Rbar*Rbar
t2bar <- trigonometric.moment(circdat, p=2, center=TRUE)
abar2 <- t2bar$cos ; bbar2 <- t2bar$sin
if (indsym == 1) {bbar2 <- 0}
div <- 2*n*Rbar2
mubc <- tbar+(bbar2/div) ; tbarstderr <- sqrt((1-abar2)/div)
```

```
if (mubc > 2*pi) {mubc <- mubc-2*pi} else
if (mubc < 0) {mubc <- mubc+2*pi}
dist <- pi-abs(pi-abs(mubc-mu0)) ; z <- dist/tbarstderr
return(list(z, mubc))
}
```

In Section 5.1.2 we rejected uniformity in favour of an underlying unimodal distribution with a mean direction corresponding to 15:00 hours for the intensive care admissions data. However, after calculating a 95% confidence interval for μ in Section 5.3.2, we rejected the null hypothesis of a population mean direction corresponding to 3pm at the 5% significance level. Here, in order to illustrate the use of the test based on (5.10) and the calculation of its p-value, we apply the function **SpecMeanTestRes** to test the null hypothesis that the population mean direction is 15:00 hours (or 3.9270 radians). Remember, once more, that in Section 5.2 we found no significant evidence against reflective symmetry for these data. Running the code:

```
cdat <- circular(fisherB1*2*pi/24) ; sym <- 1 ; mu0 <- 3.9270
testres <- SpecMeanTestRes(cdat, sym, mu0) ; z <- testres[[1]]
pval <- 2*pnorm(z, mean=0, sd=1, lower=FALSE) ; pval
```

returns a p-value of 9.1260×10^{-5}. Hence we can reject the null hypothesis of a population mean direction corresponding to 3pm at far below the 5% significance level.

For the wind directions data set, with an underlying distribution which we previously found to be significantly skewed, we can test the null hypothesis of a population mean direction corresponding to north (or 0 radians) using the commands:

```
cdat <- circular(wind) ; sym <- 0 ; mu0 <- 0
testres <- SpecMeanTestRes(cdat, sym, mu0) ; z <- testres[[1]]
pval <- 2*pnorm(z, mean=0, sd=1, lower=FALSE) ; pval
```

The p-value returned is 2.9278×10^{-10}, and hence we emphatically reject the null hypothesis.

For smaller samples, we recommend using an amended version of the bootstrap testing approach described by Fisher (1993, Section 4.4.5). This firstly involves calculating the value of the test statistic (5.10) for the original sample, z_0. Next, a mean direction shifted sample is constructed by first subtracting $\hat{\mu}_{BC}$ from each data value in the original sample and then adding μ_0 to each, so obtaining a sample with mean direction μ_0. B bootstrap samples are then drawn from the mean direction shifted sample if the underlying distribution is assumed not to be reflectively symmetric, or from the symmetrized version (around μ_0) of it (which will also have μ_0 as its mean direction) if reflective symmetry of the underlying distribution can reasonably be assumed. For each bootstrap sample the value of the test statistic (5.10) is computed, and the p-value of the test estimated by the proportion of the $(B + 1)$ z-values that are greater than or equal to z_0. The **R** function **SpecMeanTestBoot** below calculates the estimated p-value of the bootstrap version of the test for the data in the circular data object **origdat** and an hypothesized population mean direction **mu0**. Its arguments **indsym** and **B** specify whether the data are assumed to have been drawn from an underlying distribution which is reflectively symmetric (**indsym = 1**) or skewed (**indsym = 0**), and the number of bootstrap samples to be used, respectively.

```
SpecMeanTestBoot <- function(origdat, mu0, indsym, B) {
n <- length(origdat)
testres <- SpecMeanTestRes(origdat, indsym, mu0)
z <- testres[[1]] ; mubc <- testres[[2]]
shiftdat <- origdat-mubc+mu0
if (indsym == 1) {
refdat <- 2*mu0-shiftdat ; sampledat <- c(shiftdat, refdat)
} else
if (indsym == 0) { sampledat <- shiftdat }
nxtrm <- 1
for (b in 2:(B+1)) {
bootdat <- sample(sampledat, size=n, replace=TRUE)
testres <- SpecMeanTestRes(bootdat, indsym, mu0)
z[b] <- testres[[1]]
if (z[b] >= z[1]) { nxtrm <- nxtrm+1 }
}
pval <- nxtrm/(B+1) ; return(pval)
}
```

As an illustration of the use of this approach we apply it to a data set consisting of the vanishing directions of $n = 15$ homing pigeons released 16 km from their home loft. These data, measured in degrees clockwise from north, are available in the **circular** package's data object **fisherB12**. Note however, that the data in **fisherB12** were copied directly from Fisher (1993, page 245) and, as the errata list to Fisher's book (available from **http://www.valuemetrics.com.au**) identifies, the tenth observation should in fact be 185 rather than 285. The data were originally published in Schmidt-Koenig (1963). Kuiper's test rejects isotropy for these data (p-value < 0.01) but the bootstrap version of the test for reflective symmetry introduced in Section 5.2.2, with a p-value of 0.3704 when we ran it, indicates that it is reasonable to assume an underlying symmetric distribution. The compass bearing of the home loft was 149° (or 2.60 radians) from the release point, and here we make use of the **SpecMeanTestBoot** function to test the null hypothesis that the mean direction of flight is towards the home loft. We can do this using the following code:

```
x <- fisherB12 ; x <- x[-10] ; x <- c(x, 185)
cdat <- circular(x*2*pi/360) ; sym <- 1 ; mu0 <- 2.60 ; B <- 9999
pval <- SpecMeanTestBoot(cdat, mu0, sym, B) ; pval
```

The estimated p-value obtained when we ran this code was 0.1423. Using (1.1), a 95% confidence interval for the true p-value of the test is (0.135, 0.149). Hence, there is no significant evidence to reject the null hypothesis that on average the pigeons head towards their home loft. Note however that the sample mean direction is 2.94 radians (or 168°) and a nominally 95% confidence for the underlying mean direction, calculated using the bootstrap approach introduced in Section 5.3.2, is (2.53, 3.36) radians, or (145°, 192°) for the original data. This confidence interval includes the direction of the home loft (149°), along with a narrow interval of more easterly directions and a much wider interval of more southerly directions.

6

Model Fitting for a Single Sample

6.1 Introduction

In Chapter 4 we met various distributions available as models for circular data, and in Chapter 5 we saw how isotropy and reflective symmetry can be tested for. If circular uniformity is rejected, but reflective symmetry is not, **R**'s **circular** package includes functions designed specifically for carrying out maximum likelihood based inference for just three potential models: the wrapped Cauchy, wrapped normal and von Mises distributions studied in Sections 4.3.6–4.3.8. Inference for the von Mises distribution is by far the best supported. In this chapter we consider how **R** can be used to fit three particular distributions to circular data: the von Mises, Jones–Pewsey and inverse Batschelet with density (4.70). As we saw in Sections 4.3.8, 4.3.9 and 4.3.13, all three of these distributions are unimodal. The last two contain the von Mises as a special case; the Jones–Pewsey family being symmetric, whilst the inverse Batschelet family with density (4.70) is a highly flexible one containing symmetric as well as skewed distributions. Between them they offer the user a wide range of modelling capabilities. We will see how maximum likelihood based inference can be used to obtain point and interval estimates of their parameters, and how their goodness-of-fit can be investigated. For the two more flexible models we will also explore the issues of model comparison and reduction. When considering the Jones–Pewsey family, we will also show how it can be used to model grouped circular data. Given the details presented here and the functions introduced in Chapter 4, you should find it relatively easy to write your own **R** functions to fit any of the other models discussed in this book.

In order to simplify the presentation, throughout this chapter we assume that the data being analysed are represented as angles lying in $[0, 2\pi)$ measured in radians counterclockwise from zero. If the original data were not of this type they should be converted to such angles prior to applying the methods described here. If necessary, parameter estimates should be suitably back-transformed when relating them to the original data.

6.2 Fitting a von Mises Distribution

6.2.1 *Maximum Likelihood Based Point Estimation*

Using the general exploratory inferential techniques described in Chapter 5, suppose we have rejected isotropy for our data, but not symmetry. The von Mises distribution is then a potential model for our data.

As we saw in Section 4.3.8, the von Mises distribution has two parameters, its mean direction, μ, and concentration parameter, κ. Using basic calculus it is easy to show that, for a random sample drawn from a von Mises distribution, the maximum likelihood estimates of its parameters are $\hat{\mu} = \bar{\theta}$ and $\hat{\kappa} = A_1^{-1}(\bar{R})$, where $\bar{\theta}$ and $\bar{R}$ are the mean direction and mean resultant length, respectively, defined in Section 3.2, and $A_1(x) = I_1(x)/I_0(x)$, as defined in Equation (4.53). The inverse of A_1 can be computed using the **A1inv** function available within **R**'s **circular** package. That function employs a numerical root-finding method as there is no closed-form expression for the inverse of A_1.

So, maximum likelihood estimation for the von Mises distribution is fairly straightforward. However, whilst $\bar{\theta}$ is an unbiased estimator of μ, $\hat{\kappa}$ is a positively biased estimator of κ. Fisher (1993, Section 4.5.5) describes methods that can be used to correct the bias inherent in the estimation of κ, and these are implemented in the **circular** package's **mle.vonmises** function.

To illustrate how to fit the von Mises distribution using maximum likelihood estimation with bias correction, we will make use of the $n = 40$ palaeocurrent cross-bed azimuths introduced in Section 5.2.2. As we saw there, for these data Kuiper's test rejects isotropy (p-value < 0.01) but reflective symmetry is not rejected by the bootstrap version of the test of Pewsey (2002*b*) (estimated p-value $= 0.5391$). A circular data plot of the cross-bed azimuths converted to radians and represented as angles measured counterclockwise from 0 is displayed in Fig. 5.3. The commands:

```
cdat <- circular(fisherB6$set1*2*pi/360)
vMmle <- mle.vonmises(cdat, bias=TRUE)
muhat <- vMmle$mu ; semu <- vMmle$se.mu ; muhat ; semu
kaphat <- vMmle$kappa ; sekap <- vMmle$se.kappa ; kaphat ; sekap
```

return the maximum likelihood estimate $\hat{\mu} = \bar{\theta} = 3.98$ radians (or 228° clockwise from north for the original data), with a standard error of 0.28 (radians), and a bias-corrected estimate for κ of 0.83, with a standard error of 0.25. Note that the estimate of the mean direction for the assumed von Mises distribution is the same as the bias-corrected estimate of the mean direction for an assumed reflectively symmetric distribution calculated in Section 5.3.2. This is because bias$(\bar{\theta}) = 0$ for a circular distribution that is reflectively symmetric. The density for this bias-corrected maximum likelihood fit appears superimposed on the circular data plot of the data in Fig. 6.3.

6.2.2 *Confidence Interval Construction*

Given the functionality of **R**'s **circular** package, it is very easy to compute confidence intervals for the parameters of the von Mises distribution, μ and κ. Here we consider two constructions for such intervals: the first based on standard asymptotic normal theory for the maximum likelihood estimators and the second based on bootstrapping.

For the cross-bed azimuth data we can compute asymptotic normal theory confidence intervals for μ and κ with a nominal confidence level of 95% using the commands:

```
quant <- qnorm(0.975)
muint <- c(muhat-quant*semu, muhat+quant*semu)
kapint <- c(kaphat-quant*sekap, kaphat+quant*sekap)
muint ; kapint
```

In response, **R** returns the intervals $(3.43, 4.53)$ radians (or $(197°, 260°)$) for μ, and $(0.33, 1.32)$, for κ.

The bootstrap counterparts of these intervals, calculated using 10,000 bootstrap samples resampled from the original data set, can be obtained using the command:

```
mle.vonmises.bootstrap.ci(cdat, bias=TRUE, reps=10000)
```

When we ran this code, the nominally 95% confidence intervals returned were $(3.38, 4.54)$ radians (or $(194°, 260°)$), for μ, and $(0.41, 1.32)$, for κ.

Both intervals for μ are very similar to that obtained for the mean direction of an assumed underlying symmetric distribution ($(3.41, 4.55)$) in Section 5.3.2. All three intervals are relatively wide, reflecting the relatively low concentration of the data. The bootstrap interval for κ has a lower limit which is somewhat higher than that for its asymptotic normal theory counterpart. The fact that neither lower limit is very close to 0 supports the rejection of isotropy by Kuiper's test. Both intervals for κ are relatively wide, again reflecting the low concentration of the data. Here there is little difference between the intervals obtained using the two methods. More generally, in situations involving relatively few observations, the bootstrap intervals should be considered the more reliable.

Given their nominal confidence level of 95%, the intervals calculated above can be used to carry out hypothesis tests for specific values of the individual parameters at the $(100 - 95)\% = 5\%$ significance level. So, for instance, if it were hypothesized that the cross-bed azimuths were drawn from a von Mises population with mean direction $\mu = \pi$ then, as neither confidence interval for μ contains the value π, we would reject the null hypothesis at the 5% significance level. Similarly, if it were hypothesized that $\kappa = 1$ for the underlying von Mises population then, as both confidence intervals for κ contain the value 1, we would not reject that null hypothesis at the 5% significance level.

6.2.3 *Goodness-of-fit*

In the previous subsection we saw how to carry out inference under the assumption that our data form a random sample drawn from a von Mises population. Having fitted a von Mises distribution to our data, a fundamental question that arises is whether the data could have actually been drawn from that fitted distribution. This is the so-called 'goodness-of-fit' problem. Here we begin by considering graphical methods which can be used to explore the goodness-of-fit of a fitted model before moving on to formal hypothesis tests based on the application of the tests for circular uniformity discussed in Chapter 5.

Two plots are routinely used to investigate the goodness-of-fit of a posited distribution graphically: a P-P (Probability-Probability) plot and a Q-Q (Quantile-Quantile) plot. In a P-P plot, the value of the empirical distribution function for each data point is plotted against the corresponding value of the distribution function for the postulated distribution. In a Q-Q plot the values of the empirical and posited quantile functions are plotted instead. The fit of a circular distribution is judged to be good if the plotted points lie close to a diagonal line connecting the points $(0, 0)$ and $(1, 1)$ in a P-P plot, and $(0, 0)$ and $(2\pi, 2\pi)$ in a Q-Q plot. The function **vMPPQQ** below produces a combined diagram with a P-P and a Q-Q plot for a circular data object **circdat** and a postulated von Mises distribution with specified parameter values **mu** and **kappa**.

```
vMPPQQ <- function(circdat, mu, kappa) {
edf <- ecdf(circdat)
tdf <- pvonmises(circdat, mu, kappa, from=circular(0), tol = 1e-06)
tqf <- qvonmises(edf(circdat), mu, kappa, from=circular(0), tol = 1e-06)
par(mfrow=c(1,2), mai=c(0.90, 1.1, 0.05, 0.1), cex.axis=1.2, cex.lab=1.5)
plot.default(tdf, edf(circdat), pch=16, xlim=c(0,1), ylim=c(0,1), xlab = "von Mises distribution
function", ylab = "Empirical distribution function")
xlim <- c(0,1) ; ylim <- c(0,1) ; lines(xlim, ylim, lwd=2, lty=2)
plot.default(tqf, circdat, pch=16, xlim=c(0,2*pi), ylim=c(0,2*pi), xlab = "von Mises quantile
function", ylab = "Empirical quantile function")
xlim <- c(0,2*pi) ; ylim <- c(0,2*pi) ; lines(xlim, ylim, lwd=2, lty=2)
}
```

The diagram in Fig. 6.1 was produced for the von Mises distribution fitted to the cross-bed azimuths in Section 6.2.1 using the command:

```
vMPPQQ(cdat, muhat, kaphat)
```

Although most of the points displayed within the plots lie close to their respective diagonal lines, a few fall relatively far away from them. This is particularly the case for the points close

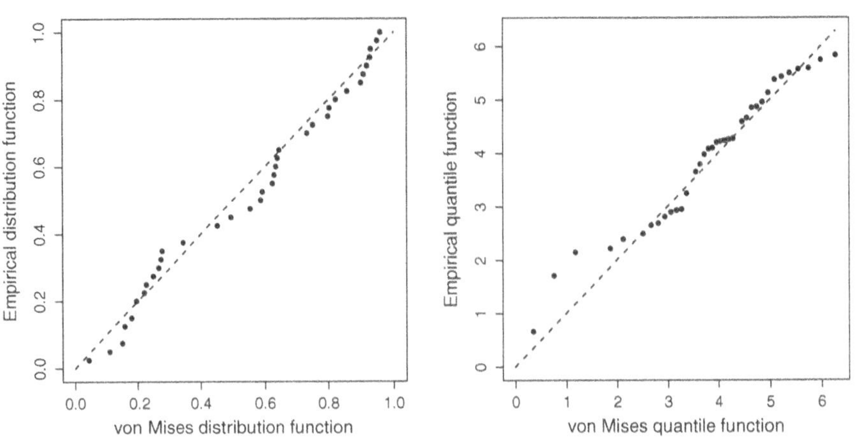

Figure 6.1 P-P plot (left) and Q-Q plot (right) for the von Mises distribution fitted to the 40 cross-bed azimuths using maximum likelihood estimation with bias correction for the estimation of κ

to the origin in the Q-Q plot, the pattern formed by them indicating a dearth of points in the arc $(0, \pi/2)$ relative to the number predicted by the fitted von Mises distribution. Clearly, the interpretation of these types of plots is rather subjective and generally we will not know just how big any deviations from their diagonal lines should be for us to conclude that there is significant evidence of lack of fit.

A more formal approach to investigating goodness-of-fit is to use hypothesis testing. Appealing to the circular analogue of the probability integral transformation, it follows implicitly that the goodness-of-fit of a posited distribution with distribution function $F(\theta)$ can be tested by calculating the values of $2\pi F(\theta_1), \ldots, 2\pi F(\theta_n)$ and applying any test of circular uniformity to them. If the data do come from the postulated distribution, then we would expect circular uniformity not to be rejected. The problem with this approach is that the usual critical values of the tests for circular uniformity do not apply if the parameters of the putative distribution have been estimated from the data. The difference between the correct critical values and those for the usual tests of circular uniformity should not be great, however, for large sample sizes. For goodness-of-fit testing for a von Mises distribution fitted using maximum likelihood estimation (without bias correction for estimation of κ), the critical values of Watson's U^2 test were obtained by Lockhart and Stephens (1985). The function **watson.test** available within **R**'s **circular** package implements this version of the test if its argument **dist** is specified as **vonmises**. Alternatively, as we shall show below, the significance of any circular uniformity test statistic employed as a goodness-of-fit statistic can be established using a parametric bootstrap approach.

To illustrate the above ideas, consider once more the cross-bed azimuth data. We can apply Watson's U^2 test using the command:

```
watson.test(cdat, dist="vonmises")
```

The p-value of the test is quoted as lying in the interval $(0.05, 0.1)$. However, it should be remembered that the test investigates the goodness-of-fit of the maximum likelihood fitted von Mises distribution without bias correction in the estimation of κ. According to Watson's U^2 test, that fit cannot be rejected as a model for the data at the 5% significance level.

Alternatively, we can apply the Kuiper, Rayleigh, Rao spacing and Watson U^2 tests for circular uniformity to the values of $2\pi F(\theta_1), \ldots, 2\pi F(\theta_n)$. This approach is implemented in the function **vMGoF**.

```
vMGoF <- function(circdat, mu, kappa) {
tdf <- pvonmises(circdat, circular(mu), kappa, from=circular(0), tol=1e-06)
cunif <- circular(2*pi*tdf)
kuires <- kuiper.test(cunif) ; rayres <- rayleigh.test(cunif)
raores <- rao.spacing.test(cunif) ; watres <- watson.test(cunif)
return(list(kuires, rayres, raores, watres))
}
```

We can run **vMGoF** for the von Mises fit to the cross-bed azimuths using the commands:

```
vMGoFRes <- vMGoF(cdat, muhat, kaphat) ; vMGoFRes
```

The p-values returned for the four tests are: > 0.15, 0.80, within $(0.05, 0.10)$ and > 0.1, respectively. Thus, none of the tests rejects circular uniformity at the 5% significance level for the values of $2\pi F(\theta_i)$, and hence the goodness-of fit of the von Mises distribution fitted to the θ_i. However, whilst these tests investigate the goodness-of-fit of the von Mises distribution fitted using maximum likelihood estimation together with bias correction, they do not allow for the fact that the parameters have been estimated.

The bootstrap approach to goodness-of-fit testing for the von Mises distribution fitted using maximum likelihood estimation together with bias correction in the estimation of κ, and which allows for the fact that the parameters have been estimated, can be implemented using the function **vMGoFBoot** below.

```
vMGoFBoot <- function(origdat, B) {
n <- length(origdat) ; vMmle <- mle.vonmises(origdat, bias=TRUE)
muhat0 <- vMmle$mu ; kaphat0 <- vMmle$kappa
tdf <- pvonmises(origdat, muhat0, kaphat0, from=circular(0), tol = 1e-06)
cunif <- circular(2*pi*tdf) ; unitest0 <- 0 ; nxtrm <- 0 ; pval <- 0
for (k in 1:4) {unitest0[k]=0 ; nxtrm[k]=1}
unitest0[1] <- kuiper.test(cunif)$statistic
unitest0[2] <- rayleigh.test(cunif)$statistic
unitest0[3] <- rao.spacing.test(cunif)$statistic
unitest0[4] <- watson.test(cunif)$statistic
for (b in 2:(B+1)) {
bootsamp <- rvonmises(n, muhat0, kaphat0)
vMmle <- mle.vonmises(bootsamp, bias=TRUE)
muhat1 <- vMmle$mu ; kaphat1 <- vMmle$kappa
tdf <- pvonmises(bootsamp, muhat1, kaphat1, from=circular(0), tol = 1e-06)
cunif <- circular(2*pi*tdf)
kuiper1 <- kuiper.test(cunif)$statistic
if (kuiper1 >= unitest0[1]) {nxtrm[1] <- nxtrm[1] + 1}
rayleigh1 <- rayleigh.test(cunif)$statistic
if (rayleigh1 >= unitest0[2]) {nxtrm[2] <- nxtrm[2] + 1}
rao1 <- rao.spacing.test(cunif)$statistic
if (rao1 >= unitest0[3]) {nxtrm[3] <- nxtrm[3] + 1}
watson1 <- watson.test(cunif)$statistic
if (watson1 >= unitest0[4]) {nxtrm[4] <- nxtrm[4] + 1}
}
for (k in 1:4) {pval[k] <- nxtrm[k]/(B+1)}
return(pval)
}
```

The function **vMGoFBoot** first fits a von Mises distribution to the original data using bias correction in the estimation of κ. The test statistic values for the four tests of uniformity are then computed for the values of $2\pi\hat{F}(\theta_1), \ldots, 2\pi\hat{F}(\theta_n)$. The whole process is then repeated for each of B parametric bootstrap samples simulated from the same von Mises distribution that was fitted to the original data. The p-value of one of the tests is then estimated by the proportion of its $(B + 1)$ test statistic values that are at least as extreme as (here, greater than or equal to) its test statistic value for the original data. When we applied the function **vMGoFBoot** to the cross-bed azimuths, using the code:

```
B <- 9999 ; pval <- vMGoFBoot(cdat, B) ; pval
```

the estimated p-values returned by the four tests were 0.12, 0.27, 0.03 and 0.12, respectively. Only the p-value for the Watson U^2 test is consistent with the p-values obtained above without any allowance for the fact that the parameters have been estimated. The p-value for the Kuiper test is lower than quoted above (> 0.15), that for the Rayleigh test is much lower than it was (0.80) and, importantly, the Rao spacing test rejects the fitted von Mises distribution as a model for the data at the 5% level. Clearly, the p-values obtained using the parametric bootstrap approach should be favoured as they allow for the parameter estimation inherent in model fitting. Three of those p-values indicate that the von Mises distribution provides a reasonable fit to the cross-bed azimuths, whilst that for the Rao spacing test suggests we should strive to identify a better model. For this reason, we will reconsider the modelling of these data in Section 6.3.1.

6.3 Fitting a Jones–Pewsey Distribution

The Jones–Pewsey family of unimodal symmetric distributions, introduced in Section 4.3.9, has three parameters: the location and concentration parameters, μ and κ, like the parameters of the von Mises distribution, and the shape parameter $-\infty < \psi < \infty$. The von Mises distribution is obtained when $\psi = 0$. In this section we consider inference for its parameters using maximum likelihood based methods of inference. We also address the issues of model comparison and reduction, goodness-of-fit, and the modelling of grouped circular data. Throughout, we make use of the **R** functions introduced in Section 4.3.9.

6.3.1 *Maximum Likelihood Point Estimation*

Generally, there are no closed-form expressions for the maximum likelihood estimates of the three parameters and numerical methods of optimization must be used to obtain them via maximization of the log-likelihood function; or, equivalently, minimization of -1 times the log-likelihood function (which we will refer to henceforth as the negative log-likelihood function). We pose the problem in these latter terms because optimization algorithms are generally written with minimization in mind. **R**'s **optim** function incorporates such a minimization algorithm. However, as well as minimizing the negative log-likelihood function, it provides us with the maximum likelihood estimates of the parameters and other useful results that can be employed in subsequent aspects of inference.

The function **JPmle** below returns the maximum value of the the log-likelihood function, the maximum likelihood estimates of μ, κ and ψ, and a numerical approximation of the Hessian matrix for a random sample assumed to have been drawn from a Jones–Pewsey population. The Hessian matrix consists of the second derivatives of the negative log-likelihood function with respect to the parameters. Its inverse is the observed Fisher information matrix. Henceforth it is assumed that the objects **lcircdat** and **lcdat** are *linear* (not *circular*) data objects containing values in $[0, 2\pi)$.

```
JPmle <- function(lcircdat) {
s <- sum(sin(lcircdat)) ; c <- sum(cos(lcircdat))
muvM <- atan2(s,c) ; if (muvM < 0) { muvM <- muvM+2*pi }
n <- length(lcircdat) ; kapvM <- A1inv(sqrt(s*s+c*c)/n)
JPnll <- function(p) {
mu <- p[1] ; kappa <- p[2] ; psi <- p[3] ; parlim <- abs(kappa*psi)
if (parlim > 10) {y <- 9999.0 ; return(y)}
else { ncon <- JPNCon(kappa, psi)
y <- -sum(log(JPPDF(lcircdat, mu, kappa, psi, ncon))) ; return(y) }
}
out <- optim(par=c(muvM, kapvM, 0), fn=JPnll, gr = NULL, method = "L-BFGS-B", lower =
c(muvM-pi, 0, -Inf), upper = c(muvM+pi, Inf, Inf), hessian=TRUE)
muhat <- out$par[1] ; kaphat <- out$par[2] ; psihat <- out$par[3]
if (muhat < 0) { muhat <- muhat+2*pi } else
if (muhat >= 2*pi) {muhat <- muhat-2*pi}
maxll <- -out$value; HessMat <- out$hessian
return(list(maxll, muhat, kaphat, psihat, HessMat))
}
```

The first four lines of **JPmle** deal with data entry and the computation of the sample size and the estimates $\hat{\mu}_{vM}$ and $\hat{\kappa}_{vM}$ for an assumed underlying von Mises distribution. The next six lines define the function **JPnll** used to compute the negative log-likelihood function subject to the constraint $|\psi\kappa| < 10$. This constraint is less stringent than the one suggested by Jones and Pewsey (2005) and is imposed to avoid overflow problems when computing values of the density and hence the negative log-likelihood. The **optim** function is then used to minimize the negative log-likelihood over the parameter space $\hat{\mu}_{vM} - \pi < \mu < \hat{\mu}_{vM} + \pi$, $\kappa > 0$ and $-\infty < \psi < \infty$ (together with the constraint referred to above), using the 'L-BFGS-B' method of optimization due to Byrd *et al.* (1995) which allows the user to impose box-constraints on the parameters. The estimates $\hat{\mu}_{vM}$ and $\hat{\kappa}_{vM}$, together with the value $\psi = 0$ corresponding to a von Mises distribution, are used as starting values. We also request the Hessian matrix be approximated numerically at the end of the optimization process. The last five lines of code before the closing curly bracket assign and output the maximized value of the log-likelihood function, the maximum likelihood estimates and the approximated Hessian matrix.

In the previous section we fitted a von Mises distribution to the cross-bed azimuth data. When investigating its goodness-of-fit, the Rao spacing test suggested a lack of fit. As an alternative model, in this section we will investigate the fit of the Jones–Pewsey family. For the cross-bed azimuths, running the code:

```
lcdat <- fisherB6$set1*2*pi/360 ; JPmleRes <- JPmle(lcdat) ; JPmleRes
```

returns a maximized value of the log-likelihood function of −64.76, the maximum likelihood estimates $\hat{\mu} = 3.97$, $\hat{\kappa} = 1.45$ and $\hat{\psi} = 1.09$, and the approximation of the Hessian matrix evaluated at the maximum likelihood solution. Note that the estimate of ψ is very close to 1, corresponding to a cardioid distribution. We will return to this point shortly.

6.3.2 *Confidence Interval Construction*

Here we consider three different constructions of confidence intervals for the parameters of a Jones–Pewsey distribution. The first is based on asymptotic normal theory for the maximum likelihood estimates and, having previously calculated the maximum likelihood estimates and the Hessian matrix using the function **JPmle** defined in Section 6.3.1, is easily applied using the function:

```
JPNTCI <- function(muest, kapest, psiest, HessMat, conflevel) {
alpha <- (100-conflevel)/100 ; quant <- qnorm(1-alpha/2)
infmat <- solve(HessMat) ; standerr <- sqrt(diag(infmat))
muint <- c(muest-quant*standerr[1], muest+quant*standerr[1])
kapint <- c(kapest-quant*standerr[2], kapest+quant*standerr[2])
psiint <- c(psiest-quant*standerr[3], psiest+quant*standerr[3])
return(list(muint, kapint, psiint))
}
```

The arguments of **JPNTCI** are the maximum likelihood estimates, the estimated Hessian matrix, and the level for the nominally $100(1-\alpha)\%$ confidence intervals to be computed. In the second line of code the value of the $(1-\alpha/2)$ quantile of the standard normal distribution is computed. Next, the Hessian matrix is inverted to obtain the observed Fisher information matrix. The elements in the diagonal of the information matrix are then stripped out and their square roots taken. These are the (asymptotic) standard errors of the maximum likelihood estimates. The final four lines of code compute and return the nominally $100(1-\alpha)\%$ confidence intervals for the individual parameters.

Continuing with our analysis of the cross-bed azimuths, the code:

```
muhat <- JPmleRes[[2]] ; kaphat <- JPmleRes[[3]] ; psihat <- JPmleRes[[4]]
HessMat <- JPmleRes[[5]] ; conflevel <- 95
jpntci <- JPNTCI(muhat, kaphat, psihat, HessMat, conflevel) ; jpntci
```

returns the nominally 95% confidence intervals $(3.45, 4.49)$ for μ, $(0.24, 2.65)$ for κ and $(0.07, 2.10)$ for ψ. The fact that the last interval does not contain the value 0 provides significant evidence at the 5% significance level that the population from which the data were drawn was not von Mises.

Our second approach to calculating confidence intervals for the individual parameters is based on the use of profile log-likelihood functions and asymptotic chi-squared theory for the distribution of the likelihood-ratio statistic. To clarify ideas, we will focus on the construction of a confidence interval for ψ. First, in order to obtain a reasonable approximation to the profile log-likelihood of ψ, we choose a sequence of ψ-values spanning an interval where we think the confidence interval might be located. The confidence interval calculated using the first construction discussed above will be useful in identifying such an exploratory interval. It is simplest if a uniform separation between the ψ-values is used. The separation should be relatively small, otherwise the approximation to the profile log-likelihood function will be overly coarse. For a specified ψ-value, the log-likelihood function is maximized over the other two parameters, μ and κ, so as to obtain the value of the profile log-likelihood of ψ for that specified ψ-value. The function **JPpllpsi** below computes values of the profile log-likelihood function of ψ for the (linear) data object **lcircdat** and the sequence of ψ-values in **psival**. Its other two arguments are the maximum likelihood estimates of μ and κ for an assumed underlying Jones–Pewsey distribution calculated previously using the function **JPmle**.

```
JPpllpsi <- function(lcircdat, muhat, kaphat, psival) {
npsival <- length(psival) ; pllpsi <- 0
for (j in 1:npsival) {
psi0 <- psival[j]
JPnllpsi0 <- function(p) {
mu <- p[1] ; kappa <- p[2] ; psi <- psi0 ; parlim <- abs(kappa*psi)
if (parlim > 10) { y <- 9999.0 ; return(y) }
else {
ncon <- JPNCon(kappa, psi)
y <- -sum(log(JPPDF(lcircdat, mu, kappa, psi, ncon))) ; return(y) }
}
out <- optim(par=c(muhat, kaphat), fn=JPnllpsi0, gr = NULL, method = "L-BFGS-B", lower =
c(muhat-pi, 0), upper = c(muhat+pi, Inf))
pllpsi[j] <- -out$value
}
return(pllpsi)
}
```

A plot of the values of the profile log-likelihood against their respective ψ-values provides an approximation to the complete profile log-likelihood function of ψ. A $100(1-\alpha)\%$ confidence interval for ψ contains those ψ-values with profile log-likelihood values greater than

$$\ell_{\max} - \frac{\chi_1^2(1-\alpha)}{2},$$

where $\ell_{\max}$ denotes the maximum value of the (full) log-likelihood function and $\chi_1^2(1-\alpha)$ denotes the $1-\alpha$ quantile of the chi-squared distribution with one degree of freedom. Such an interval contains all those ψ-values which would not be rejected at the $100\alpha\%$ significance level by a likelihood ratio test for a specified value of ψ. The function **JPpllpsiPlotCI** below produces a plot of the profile log-likelihood function of ψ with a dashed line at a height of $\chi_1^2(1-\alpha)/2$ below the maximum value of the log-likelihood function, **maxll**, superimposed on it. The arguments **psival** and **pllpsi** contain the sequence of ψ-values and their corresponding values of the profile log-likelihood computed previously using **JPpllpsi**. Finally, linear interpolation is used to compute a nominally $100(1-\alpha)\%$ confidence interval for ψ.

```
JPpllpsiPlotCI <- function(maxll, psival, pllpsi, conflevel) {
npsival <- length(psival) ; alpha <- (100-conflevel)/100
par(mai=c(0.90, 0.95, 0.05, 0.1), cex.axis=1.2, cex.lab=1.5)
plot(psival,pllpsi,type="l",lwd=2, xlab=expression(psi),ylab=expression(pll(psi)))
cutpoint <- maxll-qchisq(1-alpha, df=1)/2
xlim <- c(psival[1], psival[npsival]) ; ylim <- c(cutpoint, cutpoint)
lines(xlim, ylim, lwd=2, lty=2)
npsivalm1 <- npsival-1
for (j in 1:npsivalm1) {
if (pllpsi[j] < cutpoint) {
jp1 <- j+1
if (pllpsi[jp1] > cutpoint) {
grad <- (pllpsi[jp1]-pllpsi[j])/(psival[jp1]-psival[j])
con <- pllpsi[j]-grad*psival[j] ; psilo <- (cutpoint-con)/grad } }
if (pllpsi[j] > cutpoint) {
jp1 <- j+1
if (pllpsi[jp1] < cutpoint) {
grad <- (pllpsi[j]-pllpsi[jp1])/(psival[j]-psival[jp1])
```

```
con <- pllpsi[j]-grad*psival[j] ; psiup <- (cutpoint-con)/grad } }
}
return(list(psilo, psiup))
}
```

For the cross-bed azimuths, we can obtain a plot of the profile log-likelihood function for ψ and compute a nominally 95% confidence interval for ψ using the commands:

```
psival <- seq(-0.05, 3.3, by=0.05)
pllpsi <- JPpllpsi(lcdat, muhat, kaphat, psival)
maxll <- JPmleRes[[1]] ; conflevel <- 95
pllpsiCI <- JPpllpsiPlotCI(maxll, psival, pllpsi, conflevel) ; pllpsiCI
```

The plot produced by the last four lines of code appears in Fig. 6.2, together with analogous plots for the other two parameters computed using the functions **JPpllmu**, **JPpllmuPlotCI**, **JPpllkap** and **JPpllkapPlotCI** available from the website. The interval for ψ returned by **R** is $(-0.01, 3.10)$. According to this interval, the von Mises distribution is a potential model for the data. The analogous intervals for μ and κ, calculated from their profile log-likelihood functions displayed in Fig. 6.2, are $(3.40, 4.45)$ and, effectively, $(0.63, \infty)$. The profile log-likelihoods for μ and ψ are roughly quadratic in shape and the confidence intervals calculated from them do not differ greatly from their analogues calculated previously using asymptotic normal theory. However, the profile log-likelihood for κ is very flat, reflecting the fact that, for virtually any κ-value greater than 1, a Jones–Pewsey distribution can be identified with a log-likelihood value close to that of the maximum likelihood solution. As a consequence, the upper limits of the two intervals for κ have little in common. Between them, they provide little insight as to the potential values that κ might take.

The problem identified at the end of the last paragraph can be resolved using a third approach to confidence interval construction, based on the use of the parametric bootstrap. Being computer-intensive, this approach requires considerably more CPU time. However, for small-sized samples the results obtained using it will be more reliable than those obtained using the other two approaches considered previously, because it does not require any asymptotic theory to hold. It can be implemented using the following function.

```
JPCIBoot <- function(lcircdat, conflevel, B) {
n <- length(lcircdat) ; alpha <- (100-conflevel)/100
JPmleRes <- JPmle(lcircdat) ; muest <- JPmleRes[[2]]
kapest <- JPmleRes[[3]] ; psiest <- JPmleRes[[4]]
ncon <- JPNCon(kapest, psiest))
for (b in 2:(B+1)) {
jpdat <- JPSim(n, muest[1], kapest[1], psiest[1], ncon)
JPmleRes <- JPmle(jpdat) ; muest[b] <- JPmleRes[[2]]
kapest[b] <- JPmleRes[[3]] ; psiest[b] <- JPmleRes[[4]]
}
dist <- pi-abs(pi-abs(muest-muest[1])) ; sdist <- sort(dist)
mulo <- muest[1]-sdist[(B+1)*(1-alpha)]
muup <- muest[1]+sdist[(B+1)*(1-alpha)]
skapest <- sort(kapest)
kaplo <- skapest[(B+1)*alpha/2] ; kapup <- skapest[(B+1)*(1-alpha/2)]
spsiest <- sort(psiest)
psilo <- spsiest[(B+1)*alpha/2] ; psiup <- spsiest[(B+1)*(1-alpha/2)]
return(list(mulo, muup, kaplo, kapup, psilo, psiup))
}
```

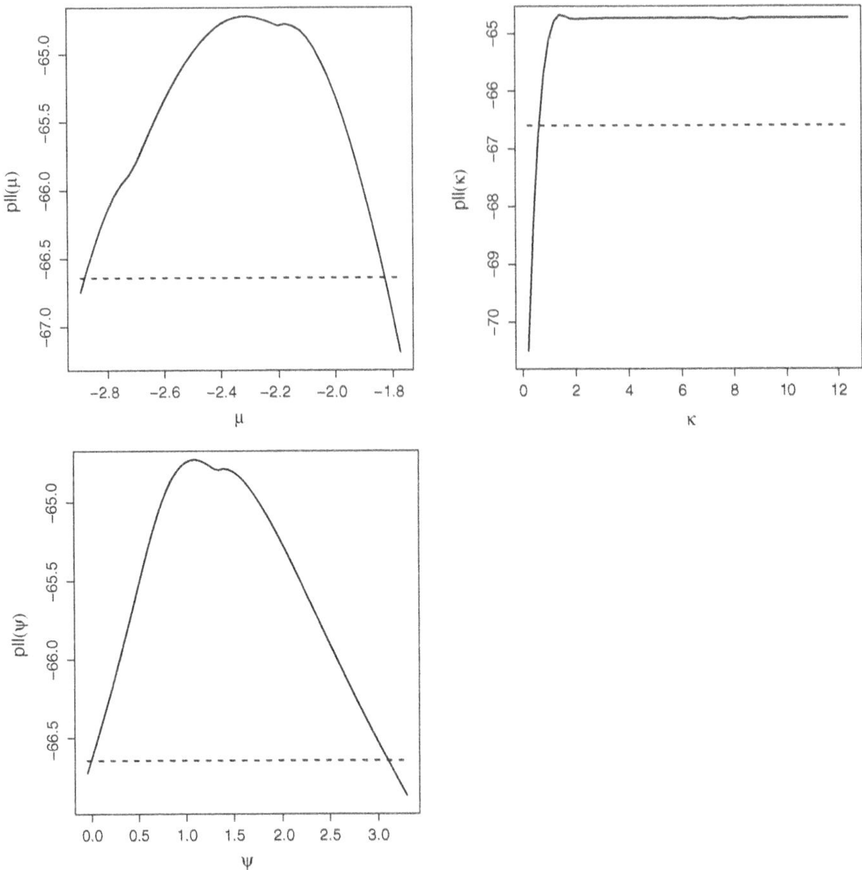

Figure 6.2 Profile log-likelihood functions for the parameters μ, κ and ψ of a Jones–Pewsey distribution fitted to the 40 cross-bed azimuths. In each panel, the dashed horizontal line lies at a height of $\frac{1}{2}\chi_1^2(0.95)$ below the maximized value of the log-likelihood −64.76

This function first calculates the maximum likelihood estimates of μ, κ and ψ for the original data set of size n. B parametric bootstrap random samples of size n are then simulated from the fitted Jones–Pewsey distribution and the maximum likelihood estimates of μ, κ and ψ calculated for each such sample. Finally, the $(B + 1)$ estimates of a parameter are ordered from smallest to largest and a 95% confidence interval for the parameter concerned calculated from its ordered estimates. As can be seen, the computation of the intervals for the linear parameters, κ and ψ, is more straightforward than that for the circular parameter μ. The latter is based on ordering the deviations of the estimates of μ about $\hat{\mu}$ calculated for the original data. It is, by construction, symmetric about the value of $\hat{\mu}$ for the original data. When we ran the commands:

```
conflevel <- 95 ; B <- 9999
JPCIBootRes <- JPCIBoot(lcdat, conflevel, B) ; JPCIBootRes
```

for the cross-bed azimuths, the confidence intervals obtained for μ, κ and ψ were $(3.47, 4.47)$, $(0.77, 12.08)$ and $(-2.21, 2.18)$, respectively. The first and last differ little from their counterparts obtained using the asymptotic theory based approaches, although the lower limit of the interval for ψ suggests that, in addition to the von Mises distribution, the wrapped Cauchy distribution, corresponding to $\psi = -1$, is also a potential model for the data. The interval for κ provides us with a clearer idea of the potential values that κ might take.

As explained above, the results obtained using the parametric bootstrap approach will be the more reliable for small-sized data sets. For larger sample sizes the three approaches should produce similar results. Confidence interval construction based on the first approach is by far the quickest. Because of potential dependencies between the parameter estimates, the construction of confidence regions for pairs of the parameters can prove enlightening. For details, see Jones and Pewsey (2005).

6.3.3 *Model Comparison and Reduction*

A fundamental idea in statistical modelling is that of *parsimony*: if two models provide equally good fits to a data set, the least complex of the two, the one with fewer parameters, is to be preferred. Consequently, we will generally be interested in comparing the fit of competing models with a view towards identifying that model with fewest parameters that provides an adequate fit to our data. Various techniques are available for comparing statistical models, with all the best-known involving maximized values of the log-likelihood function.

A formal test of the improvement in fit of a more complex model with ν_1 free parameters relative to one of its submodels with $\nu_0 < \nu_1$ free parameters can be carried out using a *likelihood ratio test*. If $\ell^0_{\max}$ denotes the maximum of the log-likelihood for the submodel and $\ell^1_{\max}$ the maximum of the log-likelihood for the more complex model, the likelihood ratio tests statistic is

$$D = -2(\ell^0_{\max} - \ell^1_{\max}). \tag{6.1}$$

Apart from when the submodel corresponds to a point on the boundary of the parameter space (Self and Liang, 1987), the sampling distribution of D is asymptotically chi-squared with $\nu_1 - \nu_0$ degrees of freedom. The test statistic D and its asymptotic chi-squared distribution formed the basis of the profile log-likelihood approach to confidence interval construction considered in Section 6.3.2. The interval constructed there for ψ indicates, for instance, that the three-parameter Jones–Pewsey model does not provide a significant improvement in fit over those for its two-parameter cardioid and von Mises submodels at the 5% significance level.

The function **JPpsi0LRT** available from the website makes use of the functions **JPnll** and **JPnllpsi0** to implement the likelihood-ratio test based on (6.1) to investigate the improvement in fit of the three-parameter Jones–Pewsey model over a particular case of it with a specified value, **psi0**, of the shape parameter ψ. Applying it to the cross-bed azimuth data, the maximum likelihood fit of the cardioid distribution ($\psi = 1$) has a maximized log-likelihood value of -64.77. When comparing this fit with that of the full Jones–Pewsey family, the value of D is 0.03 with a p-value of 0.86. Thus, according to this large-sample version of the test, the fit for the full Jones–Pewsey model does not provide a significant improvement in fit over that for its cardioid submodel. The p-value for the likelihood ratio test comparing the fit of the full Jones–Pewsey model and its von Mises ($\psi = 0$) submodel is 0.05. However, as we saw in Section 6.3.2, there is evidence that asymptotic theory produces rather inexact results for the cross-bed azimuths, for which the sample size is just 40.

Rather than rely on asymptotic results, the parametric bootstrap can be used to establish the significance of the likelihood ratio test statistic D. First the value of D is calculated for the original sample. B parametric bootstrap samples are then simulated from the hypothesized submodel with parameter values equal to those for the maximum likelihood fit of the submodel to the original data. The value of D for each bootstrap sample is calculated and, finally, the p-value of the likelihood ratio test is estimated by the proportion of the $(B + 1)$ D-values that are greater than or equal to the value of D for the original sample.

The function **JPpsi0LRTBoot** available from the website implements this bootstrap version of the likelihood ratio test to investigate the improvement in fit of the three-parameter Jones–Pewsey model over a particular case of it with a specified value of ψ. When we applied that function to the cross-bed azimuth data, the p-values obtained for the tests comparing the fits of the cardioid and von Mises distributions with that of the full Jones–Pewsey family were 0.89 and 0.15. The first is marginally higher than its large-sample counterpart, whilst the second is three times bigger than it was. In this case, then, the asymptotic versions of the tests are marginally, and quite substantially, liberal, respectively. Both pairs of p-values indicate, nevertheless, that the full model does not provide a significant improvement in fit over either submodel.

As alternatives to a formal likelihood-ratio test, we consider the use of two information criteria routinely used to compare models that are not necessarily nested one within the other. The first is Akaike's information criterion (AIC, Akaike, 1974)

$$AIC = 2\nu - 2\ell_{\max}, \tag{6.2}$$

where ν is the number of free-ranging parameters in a model and, once more, $\ell_{\max}$ denotes the maximum value of the log-likelihood function. The second is the Bayesian information criterion (BIC, Schwarz, 1978)

$$BIC = \nu \log(n) - 2\ell_{\max}, \tag{6.3}$$

where, as usual, n denotes the sample size. For either criterion, the 'best' model is that with the lowest value of the criterion. With its multiple of ν being $\log(n)$ rather than 2, the BIC penalizes parameter-heavy models more than the AIC does when the sample size is greater than 7.

The function **JPpsi0AICBIC** available from the website can be used to calculate the values of AIC and BIC for the fit of the three-parameter Jones–Pewsey model and that of its submodel with a specified value of ψ. Applying it to the cross-bed azimuths, the values of AIC for the maximum likelihood fits for the full Jones–Pewsey model and its wrapped Cauchy ($\psi = -1$), von Mises ($\psi = 0$) and cardioid ($\psi = 1$) submodels are 135.51, 139.91, 137.32

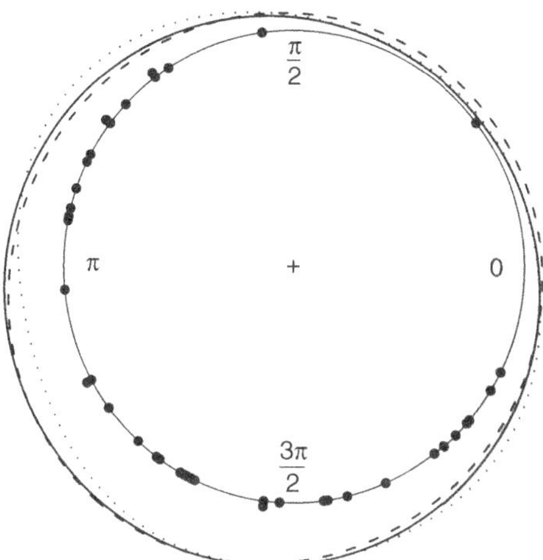

Figure 6.3 Circular data plot of the 40 cross-bed azimuths together with the densities for the cardioid (solid) and bias-corrected von Mises (dashed) fits and a kernel density estimate (dotted)

and 133.54, respectively. The corresponding values of BIC are 140.58, 143.29, 140.70 and 136.92. Thus, of the four models investigated, both criteria identify the cardioid distribution as providing the best fit to the cross-bed azimuths, and the full Jones–Pewsey family the second best fit. It would appear therefore that there is no need to fit a model as complex as the three-parameter Jones–Pewsey family to the cross-bed azimuths and that the two-parameter cardioid distribution will suffice.

R functions for fitting a Jones–Pewsey with a specified value of ψ have much in common with other functions presented previously in this section and are therefore not reproduced here. They are available from the website. Applying them to the cross-bed azimuths with $\psi = 1$, the maximum likelihood estimates of the parameters of the fitted cardioid distribution are $\hat{\mu} = 3.96$ and $\hat{\kappa} = 1.43$. The density for this cardioid fit is portrayed in Fig. 6.3 together with that of the bias-corrected von Mises fit of Section 6.2.1, a circular data plot of the cross-bed azimuths and a kernel density estimate. Although there appears to be little difference between the two fitted densities, the cardioid density generally lies closer to the kernel density estimate. Nominally 95% confidence intervals for the individual parameters, obtained using the three constructions introduced in Section 6.3.2, are: $(3.50, 4.41)$, $(3.50, 4.37)$ and $(3.52, 4.40)$ for μ, and $(0.32, 2.55)$, $(0.63, \infty)$ and $(0.71, 9.19)$ for κ. Whilst the intervals obtained for μ are very similar, the upper limits of those for κ differ greatly. For the reasons explained in Section 6.3.2, the interval for κ obtained using the parametric bootstrap, $(0.71, 9.19)$, is the more reliable.

6.3.4 *Goodness-of-fit*

The goodness-of-fit of a fitted Jones–Pewsey distribution can be investigated using the same techniques as described in Section 6.2.3. The P-P and Q-Q plots for the data in the linear data object **lcircdat** and a posited Jones–Pewsey distribution with specified parameter values of **mu**, **kappa** and **psi** can be portrayed in a combined diagram using the function **JPPPQQ** below.

```
JPPPQQ <- function(lcircdat, mu, kappa, psi) {
n <- length(lcircdat) ; ncon <- JPNCon(kappa, psi)
edf <- ecdf(lcircdat) ; tdf <- 0 ; tqf <- 0
for (j in 1:n) { tdf[j] <- JPDF(lcircdat[j], mu, kappa, psi, ncon)
tqf[j] <- JPQF(edf(lcircdat)[j], mu, kappa, psi, ncon) }
par(mfrow=c(1,2), mai=c(0.90, 1.1, 0.05, 0.1), cex.axis=1.2, cex.lab=1.5)
plot.default(tdf, edf(lcircdat), pch=16, xlim=c(0,1), ylim=c(0,1), xlab = "Jones-Pewsey distribution
function", ylab = "Empirical distribution function")
xlim <- c(0,1) ; ylim <- c(0,1) ; lines(xlim, ylim, lwd=2, lty=2)
plot.default(tqf, lcircdat, pch=16, xlim=c(0,2*pi), ylim=c(0,2*pi), xlab = "Jones-Pewsey quantile
function", ylab = "Empirical quantile function")
xlim <- c(0,2*pi) ; ylim <- c(0,2*pi) ; lines(xlim, ylim, lwd=2, lty=2)
}
```

Applying **JPPPQQ** for the cardioid fit to the cross-bed azimuths identified in the previous subsection, using the commands:

```
muhat <- 3.96 ; kaphat <- 1.43 ; psi <- 1 ; JPPPQQ(lcdat, muhat, kaphat, psi)
```

produces the P-P and Q-Q plots presented in Fig. 6.4. Comparing them with their counterparts in Fig. 6.1, it would appear that the points lie closer to the diagonal lines than they did in the previous plots, indicating, as expected, that the cardioid distribution provides a closer fit to the data than the von Mises.

The function **JPGoF** below implements the goodness-of-fit testing approach introduced in Section 6.2.3 but here for a posited Jones–Pewsey distribution. As can be appreciated, it is a simple adaptation of the function **vmGoF** introduced there.

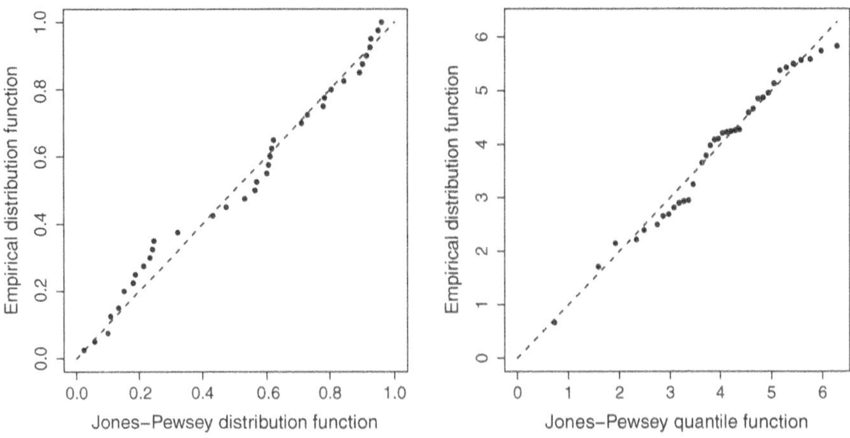

Figure 6.4 P-P plot (left) and Q-Q plot (right) for the cardioid distribution ($\psi = 1$) fitted to the 40 cross-bed azimuths using maximum likelihood estimation

```
JPGoF <- function(lcircdat, mu, kappa, psi) {
n <- length(lcircdat) ; ncon <- JPNCon(kappa, psi) ; tdf <- 0
for (j in 1:n) { tdf[j] <- JPDF(lcircdat[j], mu, kappa, psi, ncon) }
cunif <- circular(2*pi*tdf)
kuires <- kuiper.test(cunif) ; rayres <- rayleigh.test(cunif)
raores <- rao.spacing.test(cunif) ; watres <- watson.test(cunif)
return(list(kuires, rayres, raores, watres))
}
```

Running it for the cardioid fit to the cross-bed azimuths using the commands:

```
JPGoFRes <- JPGoF(cdat, muhat, kaphat, psi) ; JPGoFRes
```

returns p-values of > 0.15, 0.6015, > 0.1 and > 0.1 for the Kuiper, Rayleigh, Rao spacing and Watson tests, respectively. Thus, according to all four tests, it would appear that the fitted cardioid distribution provides an adequate fit to the data.

Remember, however, that when applying the four tests in the above way no allowance is made for the parameter estimation generally involved in fitting a distribution. The function **JPGoFBoot** below applies the parametric bootstrap versions of the four tests for a maximum likelihood fitted Jones–Pewsey distribution. An adaptation of it, **JPpsi0GoFBoot**, for testing the fit of a maximum likelihood fitted Jones–Pewsey distribution with a specified value of the shape parameter ψ, is available from the website.

```
JPGoFBoot <- function(lcircdat, B) {
n <- length(lcircdat)
JPmleRes <- JPmle(lcircdat) ; muhat0 <- JPmleRes[[2]]
kaphat0 <- JPmleRes[[3]] ; psihat0 <- JPmleRes[[4]]
ncon0 <- JPNCon(kaphat0, psihat0) ; tdf <- 0
for (j in 1:n) { tdf[j] <- JPDF(lcircdat[j], muhat0, kaphat0, psihat0, ncon0) }
cunif <- circular(2*pi*tdf) ; unitest0 <- 0 ; nxtrm <- 0 ; pval <- 0
for (k in 1:4) {unitest0[k]=0 ; nxtrm[k]=1}
unitest0[1] <- kuiper.test(cunif)$statistic
unitest0[2] <- rayleigh.test(cunif)$statistic
unitest0[3] <- rao.spacing.test(cunif)$statistic
unitest0[4] <- watson.test(cunif)$statistic
for (b in 2:(B+1)) {
bootsamp <- JPSim(n, muhat0, kaphat0, psihat0, ncon0)
JPmleRes <- JPmle(bootsamp) ; muhat1 <- JPmleRes[[2]]
kaphat1 <- JPmleRes[[3]] ; psihat1 <- JPmleRes[[4]]
ncon1 <- JPNCon(kaphat1, psihat1) ; tdf <- 0
for (j in 1:n) { tdf[j] <- JPDF(bootsamp[j], muhat1, kaphat1, psihat1, ncon1) }
cunif <- circular(2*pi*tdf)
kuiper1 <- kuiper.test(cunif)$statistic
if (kuiper1 >= unitest0[1]) {nxtrm[1] <- nxtrm[1] + 1}
rayleigh1 <- rayleigh.test(cunif)$statistic
if (rayleigh1 >= unitest0[2]) {nxtrm[2] <- nxtrm[2] + 1}
rao1 <- rao.spacing.test(cunif)$statistic
if (rao1 >= unitest0[3]) {nxtrm[3] <- nxtrm[3] + 1}
watson1 <- watson.test(cunif)$statistic
if (watson1 >= unitest0[4]) {nxtrm[4] <- nxtrm[4] + 1}
}
for (k in 1:4) {pval[k] <- nxtrm[k]/(B+1)}
return(pval)
}
```

When we ran the code:

```
psi0 <- 1 ; B <- 9999 ; pval <- JPpsi0GoFBoot(cdat, psi0, B) ; pval
```

to test the cardioid fit, the p-values for the four tests were 0.30, 0.46, 0.14 and 0.40. All four p-values are consistent with their counterparts for the tests without any allowance for parameter estimation, although the p-value for the Rayleigh test is now appreciably smaller. Neither version of any of the tests provides significant evidence of any lack of fit. It would appear, therefore, that the fitted cardioid distribution provides an adequate model for the cross-bed azimuths.

6.3.5 *Modelling Grouped Data*

As we explained in Section 5.1.1, circular data will often be grouped. With a view to fitting a Jones–Pewsey distribution to such data, consider the frequency data in Table 6.1, taken from Mardia (1972, Table 1.1). The data were obtained during an ornithological experiment in which 714 non-migratory British mallard ducks were displaced from their usual habitat. Upon release the vanishing angle of each duck was noted as belonging to one of the 18 intervals of width 20° in the table, with 0° representing north. The frequencies in the table give the number of vanishing angles in each interval. A linear histogram of the data converted to radians appears in Fig. 6.5.

More generally, for a random sample grouped into k mutually exclusive class intervals, $[\theta_{(0)}, \theta_{(1)}), [\theta_{(1)}, \theta_{(2)}), \ldots, [\theta_{(k-1)}, \theta_{(k)})$, where $\theta_{(0)} = \theta_{(k)} \pmod{2\pi}$, with n_j observations in the jth interval and thus a total of $n = n_1 + n_2 + \cdots + n_k$ observations, the log-likelihood function is given by

$$\ell = \sum_{j=1}^{k} n_j \log p_j, \tag{6.4}$$

Table 6.1 Frequencies of vanishing angles of 714 British mallard ducks.

Interval	Frequency	Interval	Frequency
$[0°, 20°)$	40	$[180°, 200°)$	3
$[20°, 40°)$	22	$[200°, 220°)$	11
$[40°, 60°)$	20	$[220°, 240°)$	22
$[60°, 80°)$	9	$[240°, 260°)$	24
$[80°, 100°)$	6	$[260°, 280°)$	58
$[100°, 120°)$	3	$[280°, 300°)$	136
$[120°, 140°)$	3	$[300°, 320°)$	138
$[140°, 160°)$	1	$[320°, 340°)$	143
$[160°, 180°)$	6	$[340°, 360°)$	69

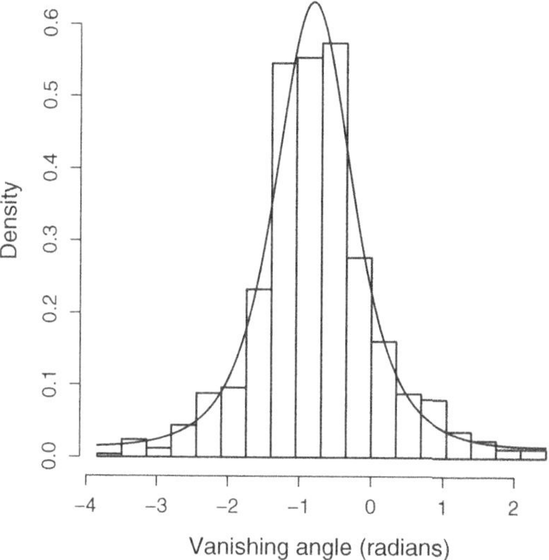

Figure 6.5 Linear histogram of the vanishing angles of 714 mallard ducks converted to radians and plotted on approximately $(\hat{\mu} - \pi, \hat{\mu} + \pi)$ together with the density of the maximum likelihood fit for the full Jones–Pewsey family

where $p_j = F(\theta_{(j)}) - F(\theta_{(j-1)})$ and F denotes the distribution function of the distribution from which the random sample is assumed to have been drawn. Fitting a Jones–Pewsey distribution to such data via maximum likelihood has much in common with what we have already seen in this section. Clearly, however, we need to define a new function to calculate the value of −1 times the log-likelihood in (6.4).

The function **JPGmlePlus** below computes maximum likelihood estimates and asymptotic normal theory based **conflevel**% confidence intervals for the parameters of a Jones–Pewsey distribution fitted to data grouped into class intervals with break points and observed frequencies specified through the argument **breaks** and **ofreqs** and with $\theta_{(0)}$ and $\theta_{(k)}$ assumed to be 0 and 2π, respectively. The values of $\ell_{\max}$, AIC and BIC for the maximum likelihood fit are also returned. The function **JPGmle** available from the website is a reduced version of **JPGmlePlus** which returns the maximum likelihood point estimates only.

```
JPGmlePlus <- function(ofreqs, breaks, conflevel) {
alpha <- (100-conflevel)/100 ; quant <- qnorm(1-alpha/2)
nfreq <- length(ofreqs) ; n <- sum(ofreqs) ; s <- 0 ; c <- 0
for (j in 1:nfreq) {
s <- s+ofreqs[j]*sin((breaks[j]+breaks[j+1])/2)
c <- c+ofreqs[j]*cos((breaks[j]+breaks[j+1])/2) }
muvM <- atan2(s,c) ; kapvM <- A1inv(sqrt(s*s+c*c)/n)
JPGnll <- function(p) {
mu <- p[1] ; kappa <- p[2] ; psi <- p[3] ; parlim <- abs(kappa*psi)
if (parlim > 10) { y <- 9999.0 ; return(y) }
```

```
else {
ncon <- JPNCon(kappa,psi) ; nll <- 0
for (j in 1:nfreq) {
pj <- JPDF(breaks[j+1],mu, kappa,psi,ncon)-JPDF(breaks[j],mu,kappa,psi,ncon)
nll <- nll - ofreqs[j]*log(pj) }
return(nll) }
}
out <- optim(par=c(muvM, kapvM, 0), fn=JPGnll, gr=NULL, method="L-BFGS-B", lower=
c(muvM-pi, 0,-Inf), upper=c(muvM+pi,Inf,Inf), hessian=T)
lmax <- -out$value ; npar <- 3
AIC <- 2*(npar-lmax) ; BIC <- (npar*log(n))-(2*lmax)
HessMat <- out$hessian ; infmat <- solve(out$hessian)
standerr <- sqrt(diag(infmat))
muhat <- out$par[1] ; kaphat <- out$par[2] ; psihat <- out$par[3]
muint <- c(muhat-quant*standerr[1], muhat+quant*standerr[1])
kapint <- c(kaphat-quant*standerr[2], kaphat+quant*standerr[2])
psiint <- c(psihat-quant*standerr[3], psihat+quant*standerr[3])
return(list(muhat, muint, kaphat, kapint, psihat, psiint, lmax, AIC, BIC))
}
```

The novel parts of this function are those dealing with calculating rough estimates of the parameters of a von Mises distribution, and the definition of the function **JPGnll** for computing the negative log-likelihood for grouped data. For the mallard duck vanishing angles, running the commands:

```
breaks <- 20*seq(from=0, to=18, by=1) ; breaks <- breaks*2*pi/360
ofreqs <- c(40, 22, 20, 9, 6, 3, 3, 1, 6, 3, 11, 22, 24, 58, 136, 138, 143, 69)
conflevel <- 95 ; JPGmleRes <- JPGmlePlus(ofreqs, breaks, conflevel)
```

returns parameter estimates for μ, κ and ψ of -0.81 radians (or 314°), 1.85 and -0.37, and nominally 95% asymptotic normal theory based confidence intervals of $(-0.87, -0.76)$ radians (or $(310^\circ, 316^\circ)$), $(1.67, 2.02)$ and $(-0.56, -0.19)$, respectively. Given the large sample size, we would not expect to obtain very different confidence intervals using the other two constructions described in Section 6.3.2. Code for computing them follows straightforwardly from that given there and above. The values of $\ell_{\max}$, AIC and BIC for the maximum likelihood fit are -1620.878, 3247.755 and 3261.468, respectively. The density corresponding to this fit appears superimposed upon a linear histogram of the vanishing angles in Fig. 6.5. Any disparities between the two appear slight. The plot was produced using the code:

```
centres <- 20*seq(from=0.5, to=17.5, by=1) ; centres <- centres*2*pi/360
gdat <- rep(centres, ofreqs) ; ciwidth <- 2*pi*20/360 ; n <- length(gdat)
for (j in 1:n) { if (gdat[j] > 7*ciwidth) {gdat[j] <- gdat[j]-2*pi} }
hbrk <- seq(from=-11*ciwidth, to=7*ciwidth, by=ciwidth)
tval <- seq(from=-11*ciwidth, to=7*ciwidth, by=0.02)
muhat <- JPGmleRes[[1]] ; kaphat <- JPGmleRes[[3]]
psihat <- JPGmleRes[[5]] ; ncon <- JPNCon(kaphat, psihat)
JPden <- JPPDF(tval, muhat, kaphat, psihat, ncon)
hist(gdat, freq=F, breaks=hbrk, main=" ", ylim=c(0,0.65), xlab="Vanishing angle (radians)", ylab=
"Density")
lines(tval, JPden)
```

The confidence interval for ψ quoted above contains neither the value 0 (von Mises) nor -1 (wrapped Cauchy), corresponding to the closest recognizable special cases. To investigate their fits we can use the function **JPGpsi0mlePlus** available from the website. It performs calculations analogous to those implemented in **JPGmlePlus** but for a Jones–Pewsey submodel with a specified value of ψ. Using it, the two special cases have maximized log-likelihood values of −1629.43 and 1638.37, AIC values of 3262.87 and 3280.75, and BIC values of 3272.01 and 3289.89. The p-values of asymptotic chi-squared theory based likelihood ratio tests for the improvement in fit of the three-parameter model over the two two-parameter submodels are 3.5×10^{-5} and 3.3×10^{-9}, respectively. According to both tests, then, the fitted Jones–Pewsey distribution provides a highly significant improvement in fit over either submodel. The AIC and BIC values support its superiority.

As in Section 6.3.4, goodness-of-fit can be explored informally using P-P and Q-Q plots. The function **JPGPPQQ** below produces a combined diagram containing both plots for grouped circular data formatted as described previously.

```
JPGPPQQ <- function(ofreqs, breaks, mu, kappa, psi) {
n <- sum(ofreqs) ; ncon <- JPNCon(kappa, psi) ; uplims <- breaks[-1]
nlims <- length(uplims) ; edf <- cumsum(ofreqs/n) ; tdf <- 0 ; tqf <- 0
for (j in 1:nlims) { tdf[j] <- JPDF(uplims[j], mu, kappa, psi, ncon)
tqf[j] <- JPQF(edf[j], mu, kappa, psi, ncon) }
par(mfrow=c(1,2), mai=c(0.90, 1.1, 0.05, 0.1), cex.axis=1.2, cex.lab=1.5)
plot.default(tdf, edf, pch=16, xlim=c(0,1), ylim=c(0,1), xlab = "Jones-Pewsey distribution function",
ylab = "Empirical distribution function")
xlim <- c(0,1) ; ylim <- c(0,1) ; lines(xlim, ylim, lwd=2, lty=2)
plot.default(tqf,uplims, pch=16, xlim=c(0,2*pi), ylim=c(0,2*pi), xlab = "Jones-Pewsey quantile
function", ylab = "Empirical quantile function")
xlim <- c(0,2*pi) ; ylim <- c(0,2*pi) ; lines(xlim, ylim, lwd=2, lty=2)
}
```

For the mallard duck vanishing angles, running the commands:

```
mu <- 5.47 ; kappa <- 1.85 ; psi <- -0.37
JPGPPQQ(ofreqs, breaks, mu, kappa, psi)
```

produces the diagram presented in Fig. 6.6. All of the points in the P-P plot lie close to the diagonal line. The Q-Q plot manifests some larger disparities due to there being more vanishing angles than predicted by the fitted Jones–Pewsey distribution in the arc ranging between, approximately, 1.5 and 3.5 radians (or $(85^\circ, 200^\circ)$).

More formally we can apply a goodness-of-fit test. However, as we are now dealing with grouped data, we can no longer apply the approach described in Sections 6.2.3 and 6.3.4 based on the different tests of continuous circular uniformity. Instead we use the parametric analogue of the approach based on bootstrapping the U_G^2 statistic introduced in Section 5.1.1 to establish the significance of the observed value of U_G^2 when the parameters of an hypothesized Jones–Pewsey distribution have been estimated from the data. First, the maximum likelihood method is used to fit a Jones–Pewsey distribution to the grouped data set under consideration, and the value of U_G^2 calculated for this fit. Then, for each of B parametric bootstrap samples simulated from the Jones–Pewsey distribution fitted to the original data set, the data are grouped into the same class intervals as those for the original data, a Jones–Pewsey distribution is fitted by maximum likelihood to the resulting grouped data and the value of U_G^2 is calculated. The p-value of the test is estimated by the proportion of the $(B+1)$ U_G^2-values that are greater than or equal to the value of U_G^2 for

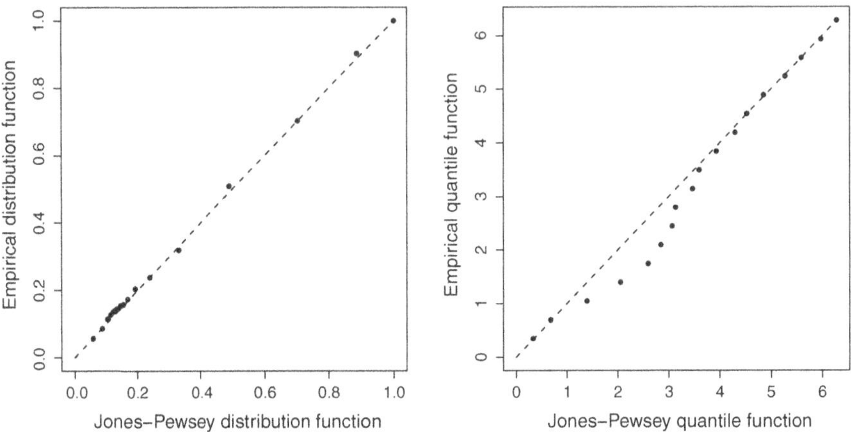

Figure 6.6 P-P plot (left) and Q-Q plot (right) for the Jones–Pewsey distribution fitted to the mallard duck vanishing angles using maximum likelihood estimation

the original data. The function **JPGUGsqBoot** below implements this parametric bootstrap testing approach. It calls the function **JPGUGsq** to calculate the value of U_G^2 for a Jones-Pewsey distribution with parameter values **mu**, **kappa** and **psi** and grouped circular data formatted as described previously.

```
JPGUGsq <- function(ofreqs, breaks, mu, kappa, psi, ncon) {
Fval <- 0 ; Pval <- 0 ; Eval <- 0 ; Dval <- 0
nbreak <- length(breaks) ; n <- sum(ofreqs)
for (j in 2:nbreak) { jm1 <- j-1
Fval[j] <- JPDF(breaks[j], mu, kappa, psi, ncon)
Pval[jm1] <- (Fval[j]-Fval[jm1]) ; Eval[jm1] <- Pval[jm1]*n
Dval[jm1] <- ofreqs[jm1]-Eval[jm1] }
Sval <- cumsum(Dval) ; Sbar <- sum(Pval*Sval)
UGsq <- sum((Sval-Sbar)*(Sval-Sbar)*Pval)/n ; return(UGsq)
}

JPGUGsqBoot <- function(ofreqs, breaks, B) {
nfreq <- length(ofreqs) ; n <- sum(ofreqs)
JPGmleRes <- JPGmle(ofreqs, breaks) ; muhat0 <- JPGmleRes[[1]]
kaphat0 <- JPGmleRes[[2]] ; psihat0 <- JPGmleRes[[3]]
ncon0 <- JPNCon(kaphat0, psihat0)
stat0 <- JPGUGsq(ofreqs, breaks, muhat0, kaphat0, psihat0, ncon0)
nxtrm <- 1
for (b in 2:(B+1)) {
freqs <- seq(1:nfreq)*0
x <- JPSim(n, muhat0, kaphat0, psihat0, ncon0)
for (j in 1:n) { for (k in 1:nfreq) {
if (x[j] >= breaks[k]) { if (x[j] < breaks[k+1]) {freqs[k] <- freqs[k]+1} } } }
JPGmleRes <- JPGmle(freqs, breaks) ; muhat1 <- JPGmleRes[[1]]
kaphat1 <- JPGmleRes[[2]] ; psihat1 <- JPGmleRes[[3]]
ncon1 <- JPNCon(kaphat1, psihat1)
stat1 <- JPGUGsq(freqs, breaks, muhat1, kaphat1, psihat1, ncon1)
if (stat1 >= stat0) {nxtrm <- nxtrm + 1}
}
pval <- nxtrm/(B+1) ; return(pval)
}
```

When we applied **JPGUGsqBoot** to the mallard duck vanishing angles, using the commands:

```
B <- 9999 ; pval <- JPGUGsqBoot(ofreqs, breaks, B) ; pval
```

the estimated p-value returned was 0.0716. Applying (1.1), a 95% confidence interval for the true p-value is (0.0665, 0.0767). Thus the test does not provide significant evidence of any lack of fit at the 5% significance level. The function **JPGpsi0UGsqBoot** available from the website implements the parametric bootstrap testing approach for a maximum likelihood fitted Jones–Pewsey submodel with a pre-specified ψ-value. Running it with the mallard duck vanishing angles, the p-values returned for the goodness-of-fit of the von Mises and wrapped Cauchy distributions were both 0.0001. Thus, the fits of both submodels are emphatically rejected.

The mallard data analysed here were also used as an illustrative example by Jones and Pewsey (2005). It should be noted that the results we have presented differ, sometimes considerably, from those reported in Jones and Pewsey (2005) where no allowance for grouping was made. Examples of maximum likelihood based inference involving other models for grouped circular data can be found in Jones and Pewsey (2012) and Abe *et al.* (2013).

6.4 Fitting an Inverse Batschelet Distribution

In the previous two sections we saw how to fit the von Mises and Jones–Pewsey distributions to unimodal circular data. Remember, both of those distributions are reflectively symmetric. Suppose now that, in an initial analysis of our data using the exploratory techniques described in Chapters 2, 3 and 5, the data appear to have been drawn from a unimodal distribution that is asymmetric. In Chapter 4 we met various skew models which we might contemplate fitting to our data. However, as explained there, by far the most flexible models are those considered in Section 4.3.13 obtained using inverse Batschelet transformation. In this final section of the chapter we show how the extension of the von Mises distribution model with density (4.70) can be fitted to unimodal circular data suspected to be skew.

As in the previous two sections, we consider point and interval estimation for the model's parameters, model comparison and reduction and goodness-of-fit. To perform these various forms of inference we make use of **R**'s **circular** package as well as functions introduced in Section 4.3.13, others introduced within the text, and others available from the website. Throughout we assume the circular data under consideration to be continuous. Grouped circular data can be analysed using methods analogous to those described in Section 6.3.5.

To illustrate the application of the methodology within **R**, we make use of a data set analysed in Jones and Pewsey (2012) on the changes in direction of a fruit fly larva. The larva was observed for a period of three minutes as it wriggled upon a flat surface, its direction of movement being noted every second. The resulting 180 changes in direction, measured in radians, are reproduced in the file **larva.dat** available from the website. They are also available as the data object **fflarvacd** within the **CircStatsInR** workspace. As Jones and Pewsey (2012) report, the form of the circular autocorrelation function is consistent with the

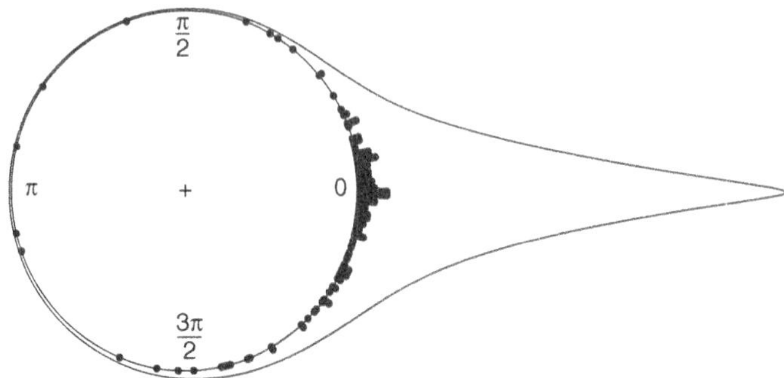

Figure 6.7 Circular data plot of the 180 changes in direction of a fruit fly larva together with the fit of the inverse Batschelet distribution with density (4.70)

changes in direction being independent, and here we make the perhaps biologically questionable assumption that they are indeed so. Copying the data file to a suitable directory on your computer, and identifying that directory as the one from which data should be read by **R**, the changes in direction can be loaded and copied in the correct format to the linear data object **lcdat** and the circular data object **cdat** using the commands:

```
angdat <- read.table(file="larva.dat", header=TRUE) ; attach(angdat)
lcdat <- changeang ; cdat <- circular(lcdat) ; n <- length(lcdat)
for (j in 1:n) {if (lcdat[j] <= 0) {lcdat[j] <- lcdat[j]+2*pi} }
```

For these data, the p-values obtained for Kuiper's test for isotropy and Pewsey's test for circular reflective symmetry, introduced in Sections 5.1.1 and 5.2, are < 0.01 and 0.06, respectively. Thus, isotropy is rejected emphatically and circular reflective symmetry appears questionable. In addition, the circular data plot of the changes in direction portrayed in Fig. 6.7 suggests the underlying distribution to be unimodal.

6.4.1 *Maximum Likelihood Point Estimation*

Generally there are no closed-form expressions for the maximum likelihood estimates of the parameters of the distribution with density (4.70) and so numerical methods of optimization must be used to obtain them via minimization of the negative log-likelihood function. Once again, **R**'s **optim** function can be used to carry out the optimization and compute various useful summaries for use in later calculations. The function **invBmle** below fits an inverse Batschelet distribution with density (4.70) to the data in the linear data object **lcircdat** and returns the maximized value of the log-likelihood, the values of AIC and BIC, the maximum likelihood estimates of ξ, κ, ν and λ, and the Hessian matrix.

```
invBmle <- function(lcircdat) {
s <- sum(sin(lcircdat)) ; c <- sum(cos(lcircdat))
xivM <- atan2(s,c) ; n <- length(lcircdat) ; kapvM <- A1inv(sqrt(s*s+c*c)/n)
invBnll <- function(p) {
xi <- p[1] ; kappa <- p[2] ; nu <- p[3] ; lambda <- p[4]
```

```
ncon <- invBNCon(kappa, lambda) ; sum <- 0
for (j in 1:n) {
sum <- sum-log(invBPDF(lcircdat[j], xi, kappa, nu, lambda, ncon)) }
return(sum)
}
out <- optim(par=c(xivM,kapvM,0.01,0.01), fn=invBnll, gr=NULL, method="L-BFGS-B",
lower=c(-pi,0,-1,-1), upper=c(pi,Inf,1,1), hessian=T)
xihat <- out$par[1] ; kaphat <- out$par[2] ; nuhat <- out$par[3]
lamhat <- out$par[4] ; maxll <- -out$value ; HessMat <- out$hessian
npar <- 4 ; AIC <- 2*(npar-maxll) ; BIC <- (npar*log(n))-(2*maxll)
return(list(maxll, AIC, BIC, xihat, kaphat, nuhat, lamhat, HessMat))
}
```

This function has much in common with the function **JPmle** used to fit the Jones–Pewsey family in Section 6.3.1, the main difference being the minimization of the negative log-likelihood function **invBnll** rather than **JPnll**. Applying it to the fruit fly larva changes in direction using the commands:

```
invBmleRes <- invBmle(lcdat) ; invBmleRes
```

the estimates obtained for the four parameters of the distribution are $\hat{\xi} = -1.05$, $\hat{\kappa} = 2.77$, $\hat{\nu} = -0.52$ and $\hat{\lambda} = 0.92$. The maximum likelihood estimate of the modal direction, $\xi - 2\nu$, is thus $\hat{\xi} - 2\hat{\nu} = -0.01$, very close to 0. The density corresponding to this fit is superimposed upon the circular data plot of the changes in direction in Fig. 6.7, generated using the commands:

```
xihat <- invBmleRes[[4]] ; kaphat <- invBmleRes[[5]]
nuhat <- invBmleRes[[6]] ; lamhat <- invBmleRes[[7]]
ncon <- invBNCon(kaphat, lamhat)
theta <- circular(seq(0, 2*pi, by=pi/3600)) ; y <- 0 ; nt <- length(theta)
for (j in 1:nt) { y[j] <- invBPDF(theta[j], xihat, kaphat, nuhat, lamhat, ncon) }
par(mai=c(0, 0, 0, 0))
plot(cdat, xlim=c(-0.8,3.3), pch=16, col="black", stack=TRUE, bins=360, cex=0.7)
lines(theta, y, lty=1, lwd=1)
```

Consistent with the estimates of ν and λ, the fitted density decays away from the modal direction more slowly in the clockwise direction (i.e. is negatively skew) and is highly peaked.

6.4.2 *Confidence Interval Construction*

The same three constructions introduced in Section 6.3.2 can be used to obtain confidence intervals for the four individual parameters of the distribution with density (4.70). The function **invBNTCI** available from the website calculates nominally $100(1-\alpha)\%$ confidence intervals for the individual parameters based on asymptotic normal theory. For the fruit fly larva changes in directions, the confidence intervals returned using the commands:

```
HessMat <- invBmleRes[[6]] ; conflevel <- 95
invbntci <- invBNTCI(xihat, kaphat, nuhat, lamhat, HessMat, conflevel)
```

are (–1.76, –0.34) for ξ, (2.46, 3.08) for κ, (–0.88, –0.17) for ν and (0.58, 1.26) for λ. The last interval clearly does not respect the constraint that $-1 \leq \lambda \leq 1$, and it is natural to truncate it to (0.58, 1). The fact that the interval for ν does not contain the value 0 supports our initial impression of an underlying distribution that is skew. Given that the interval for λ contains only positive values, there is strong evidence that the underlying distribution is more peaked than the von Mises.

Functions to implement the profile log-likelihood based approach to confidence interval construction are also available from the website. Using them with the changes in direction of the fruit fly larva produced the plots making up the four panels of Fig. 6.8 and nominally 95% confidence intervals of (–1.66, –0.36) for ξ, (2.47, 3.13) for κ, (–0.83, –0.18) for

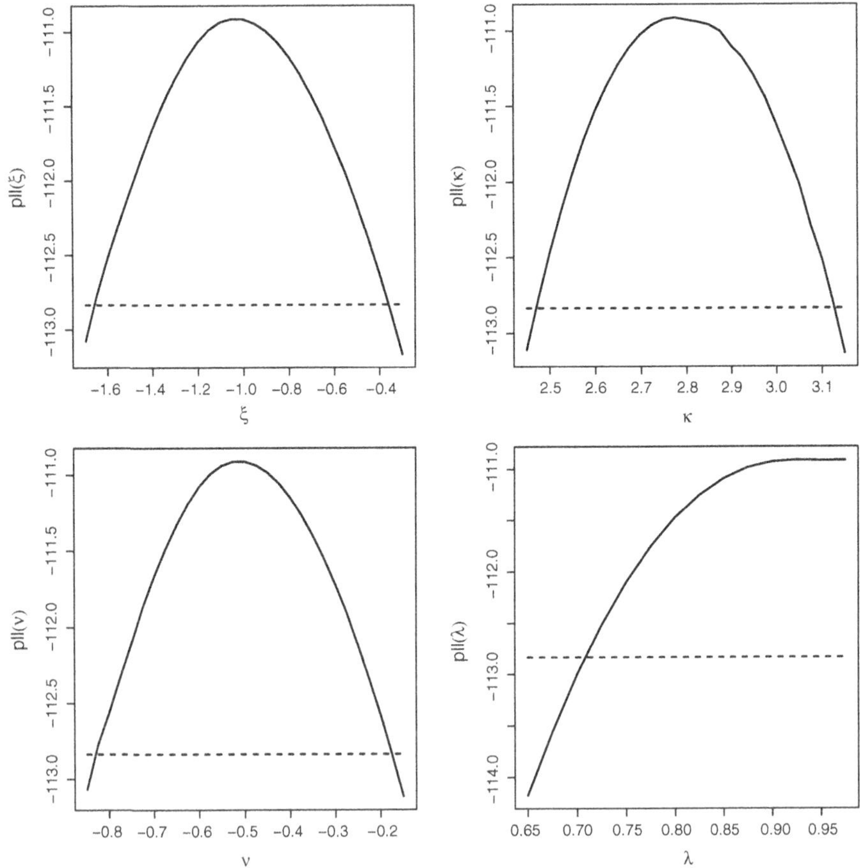

Figure 6.8 Profile log-likelihood functions for ξ (top left), κ (top right), ν (bottom left) and λ (bottom right) for the inverse Batschelet distribution with density (4.70) fitted to the 180 changes in direction of the fruit fly larva. In each panel, the dashed horizontal line lies at a height of $\frac{1}{2}\chi_1^2(0.95)$ below the maximized value of the log-likelihood –110.92

ν and $(0.71, 1)$ for λ. All four intervals are very similar to their asymptotic normal theory counterparts referred to above. The analogous interval for the mode, $\xi - 2\nu$, is calculated to be $(-0.03, 0.03)$.

The function **invBCIBoot** for computing parametric bootstrap based confidence intervals for the individual parameters and the mode is available from the website. Its structure has much in common with the function **JPCIBoot** presented in Section 6.3.2. Using it with $B = 999$ (rather than $B = 9999$) parametric bootstrap samples drawn from the distribution fitted to the fruit fly larva changes in direction, returned nominally 95% confidence intervals of $(-1.58, -0.52)$ for ξ, $(2.52, 3.16)$ for κ, $(-0.78, -0.25)$ for ν and $(0.73, 1)$ for λ. All four intervals are very similar to those calculated using the other two methods. The interval obtained for the mode, $(-0.04, 0.02)$, is also very close to its profile log-likelihood based counterpart. Thus, for these data, with a sample size as large as $n = 180$, there is little to recommend the hours of CPU time required by the parametric bootstrap approach instead of the seconds needed to implement the asymptotic normal theory and profile log-likelihood approaches.

6.4.3 *Model Comparison and Reduction*

Model comparison and reduction for the inverse Batschelet distribution with density (4.70) can also be based on likelihood ratio testing and the values taken by the AIC and BIC information criteria, as described in Section 6.3.3. When investigating model reduction there are three submodels that one might generally consider: the von Mises $(\nu = 0, \lambda = 0)$, symmetric $(\nu = 0)$, and skew-von Mises $(\lambda = 0)$ distributions. Using the functions **invBn0l0mle, invBnu0mle** and **invBlam0mle** available from the website it is simple to fit these various submodels, perform likelihood ratio tests for them and calculate their AIC and BIC values. The relevant results for the fruit fly larva changes in direction are presented in Table 6.2. The p-values of likelihood ratio tests for the improvement in

Table 6.2 Parameter estimates for the fits to the changes in direction of the fruit fly larva of, reading from right to left, the full four-parameter family with density (4.70) and its skew-von Mises $(\lambda = 0)$, symmetric $(\nu = 0)$ and von Mises $(\nu = 0, \lambda = 0)$ submodels. The maximized log-likelihood (ℓ_{max}), AIC and BIC values are also presented.

Parameter	Model			
	von Mises	Symmetric	Skew-von Mises	Full family
ξ	-0.10	-0.01	-0.95	-1.05
κ	3.67	2.77	3.84	2.77
ν	0	0	-0.47	-0.52
λ	0	0.98	0	0.92
ℓ_{max}	-155.3	-115.1	-150.1	-110.9
AIC	314.6	236.2	306.1	229.8
BIC	321.0	245.8	315.7	242.6

fit of the full four-parameter model over the skew-von Mises, symmetric and von Mises submodels are, to three decimal places, 0.000, 0.004 and 0.000, respectively. Thus the full four-parameter model is identified as providing a significant improvement in fit over all three submodels. The AIC and BIC values also identify the full model as providing the best fit to the data. Thus, for the fruit fly larva changes in direction, no reduction in the complexity of the full four-parameter model appears justified.

6.4.4 Goodness-of-fit

The goodness-of-fit of a postulated inverse Batschelet distribution with density (4.70) can be explored using P-P and Q-Q plots and formal tests of isotropy for the values of $2\pi F(\theta_1), \ldots, 2\pi F(\theta_n)$, along analogous lines to those described in Sections 6.2.3 and 6.3.4. Here we consider the goodness-of-fit of the four-parameter maximum likelihood fit to the fruit fly larva changes in direction.

Making use of the function **invBPPQQ** available from the website, the composite diagram in Fig. 6.9 was produced using the command:

```
invBPPQQ(lcdat, xihat, kaphat, nuhat, lamhat)
```

The P-P plot provides very little evidence of any lack of fit, the plotted points generally nestling tightly around the diagonal straight line. In the Q-Q plot, seven points lie relatively far below the reference line. These points correspond to changes of direction within, approximately, the arc $(1.5, 4)$ positioned around the antimode. As they are located below the diagonal line they provide evidence that the number of data values in that arc is somewhat higher than predicted by the fitted inverse Batschelet distribution.

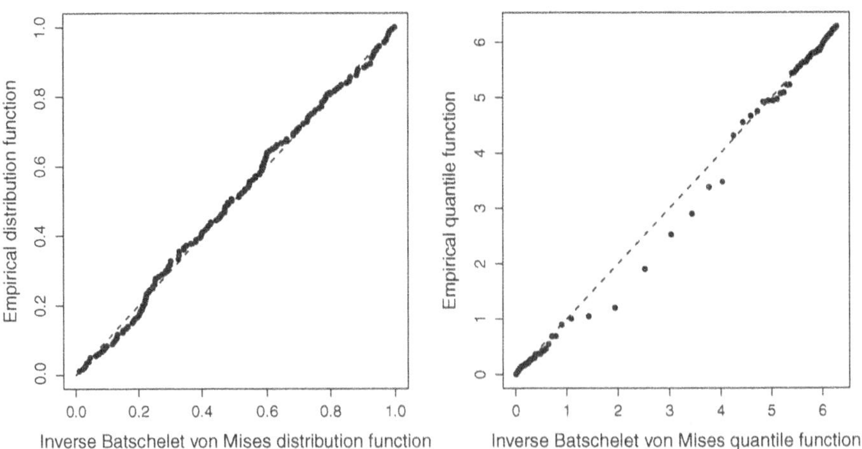

Figure 6.9 P-P plot (left) and Q-Q plot (right) for the fit of the full four-parameter inverse Batschelet distribution with density (4.70) to the 180 changes in direction of the fruit fly larva

More formally, the goodness-of-fit of a putative distribution with density (4.70) and distribution function $F(\theta)$ can be tested for by applying the Kuiper, Rayleigh, Rao spacing and Watson tests of uniformity to the values of $2\pi F(\theta_1), \ldots, 2\pi F(\theta_n)$. This approach has been implemented, without any allowance for potential parameter estimation in the identification of the posited distribution, in the function **invBGoF** available from the website. Applying it to test the goodness-of-fit of the maximum likelihood fit to the fruit fly larva changes in direction using the command:

```
invBGoF(lcdat, xihat, kaphat, nuhat, lamhat)
```

the p-values returned for the four tests are: > 0.15, 0.7155, > 0.1 and > 0.1, respectively. Given the relatively large sample size of 180 for these data, we would not expect the parametric bootstrap counterparts of these p-values, which would allow for the inherent parameter estimation, to be very different. Clearly, none of the tests provides significant evidence of any lack of fit of the fitted inverse Batschelet density.

7

Comparing Two or More Samples of Circular Data

In this chapter we explore statistical methods for use with two or more samples of circular data. We begin by discussing graphical summaries that can be used to compare the data distributions of two or more samples. We then proceed to a consideration of formal tests of the hypotheses of a common mean direction, a common median direction, a common concentration and a common distribution. In practice, we will generally *apply* these tests in the reverse order to which we have presented them. The chapter ends with the details of Moore's test for use with paired samples. Methods for estimating common measures for two or more distributions are discussed in Fisher (1993, Sections 5.3–5.4).

Formally, we assume we are comparing g independent samples, $\boldsymbol{\theta}_1, \ldots, \boldsymbol{\theta}_g$, where $\boldsymbol{\theta}_k = (\theta_{k1}, \ldots, \theta_{kn_k})^T$ is the kth sample, $k = 1, \ldots, g$. The combined sample, $\boldsymbol{\theta} = (\boldsymbol{\theta}_1^T, \ldots, \boldsymbol{\theta}_g^T)^T$, contains a total number of $N = n_1 + \cdots + n_g$ observations.

7.1 Exploratory Graphical Comparison of Samples

7.1.1 *Multiple Raw Circular Data Plot*

Throughout this chapter we will illustrate the use of the different techniques introduced using a data set that we met for the first time in Section 2.2. The data come from an experiment described by Wehner and Müller (1985) and represent the walking directions of three groups of long-legged desert ants under three different conditions. The directions, measured clockwise in degrees, are reproduced in Fisher (1993, Appendix B.10) and are available as the linear data object **fisherB10** and the circular data object **fisherB10c** that come with the **circular** library. Set 1 consists of the directions for a control group of 11 ants that had one of their eyes trained to learn the home (zero) direction and then had that eye covered and the other eye uncovered. Set 2 contains the directions for 32 ants in a first treatment group that had their naive eyes covered while being trained, and set 3 the directions of 18 ants in a second treatment group that had their naive eyes covered throughout the experiment. Raw circular data plots of the walking directions of the three groups of

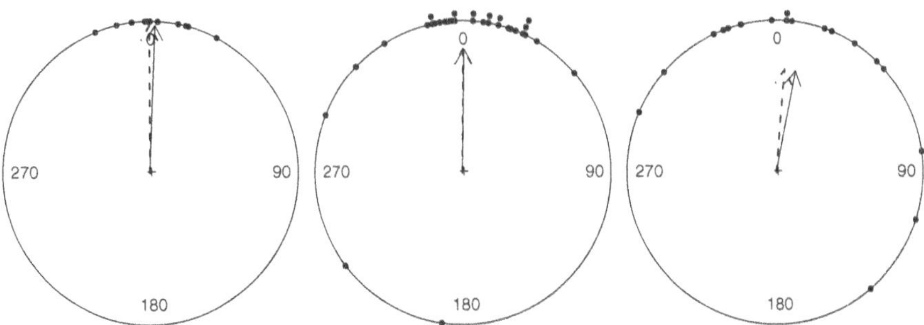

Figure 7.1 Circular data plots of the walking directions of the 11 long-legged desert ants in the control group (left), the 32 ants in the first treatment group (centre) and the 18 ants in the second treatment group (right). The arrows indicate the mean (solid) and median (dashed) directions and the mean resultant length (both)

ants are presented in Fig. 7.1. The two arrows in each plot represent the mean and median directions and the mean resultant length. The mean and median directions are similar for the control group and the first treatment group, and both are close to the zero direction. For the second treatment group the difference between the mean and median directions is greater, and both correspond to directions slightly clockwise from zero. The directions for the control group are tightly clustered around the zero direction while those for the two treatment groups are more disperse. There is perhaps some evidence of skewness in the distributions of the directions for the two treatment groups. Nevertheless, the bootstrap test introduced in Section 5.2.2 does not reject reflective symmetry for any of the data sets (estimated p-values of 0.9511, 0.1232 and 0.1691, respectively, when we ran them with the function **RSTestBoot**).

Extensions of the form of presentation employed in Fig. 7.1 can be used to compare the main distributional features of more than three samples. However, the visual comparison of multiple plots becomes increasingly complicated as the number of circular data plots included within them increases.

7.1.2 *Angular Q-Q Plot*

The angular Q-Q plot is proposed by Fisher (1993, Section 5.2) as an alternative graphical means of comparing the distributional *shapes* of two samples; but not any differences in their *central directions*, as both samples are centred around their median directions in the construction of the plot. The **R** function **TwoSampleQQ** below produces such a plot. It uses the data set with the smaller sample size as the reference distribution, and the first data set if both are of equal size. Here, **cdat1** and **cdat2** are assumed to be circular data objects containing angles measured counterclockwise from the mathematical zero direction in radians in $(-\pi, \pi]$.

```
TwoSampleQQ <- function(cdat1, cdat2) {
n1 <- length(cdat1) ; n2 <- length(cdat2)
nmin <- min(n1,n2) ; nmax <- max(n1,n2)
```

```
cdatref <- cdat1 ; cdatoth <- cdat2
if (n2 < n1) { cdatref <- cdat2 ; cdatoth <- cdat1 }
zref <- sin(0.5*(cdatref-medianCircular(cdatref))) ; szref <- sort(zref)
zoth <- sin(0.5*(cdatoth-medianCircular(cdatoth))) ; szoth <- sort(zoth)
koth <- 0 ; szothred <- 0 ; szreffin <- 0
for (j in 1:nmin) { koth[j] <- 1+nmax*(j-0.5)/nmin
szothred[j] <- szoth[koth[j]] ; szreffin[j] <- szref[j] }
par(mai=c(0.90, 0.9, 0.05, 0.1), cex.axis=1.2, cex.lab=1.5)
plot(szreffin, szothred, pch=16, xlim=c(-1,1), ylim=c(-1,1), xlab="Smaller sample", ylab=
"Larger sample")
xlim <- c(-1,1) ; ylim <- c(-1,1) ; lines(xlim, ylim, lwd=2, lty=2)
}
```

For the walking directions of the three groups of ants, running the commands:

```
cdat1 <- circular(fisherB10$set1*2*pi/360)
cdat2 <- circular(fisherB10$set2*2*pi/360) ; TwoSampleQQ(cdat1, cdat2)
cdat3 <- circular(fisherB10$set3*2*pi/360) ; TwoSampleQQ(cdat1, cdat3)
TwoSampleQQ(cdat2, cdat3)
```

produces the angular Q-Q plots in Fig. 7.2. They appear to suggest that the shapes of the distributions for the control and first treatment groups are the most similar, at least in the vicinity of the origin, $(0, 0)$, where the two sample median directions coincide. The plots for the other two group combinations reflect differences between the distributions of their walking directions which may, or may not, be significant. As can be seen from Fig. 7.2, the number of points appearing in an angular Q-Q plot is equal to the size of the smaller of the two samples. Thus, not all of the information contained in the two samples is fully represented. It is therefore debatable whether, for pairs of data sets with substantially different sample sizes, such plots provide any real insight beyond that gleaned from a comparison of pairs of raw circular data plots, such as those in Fig. 7.1.

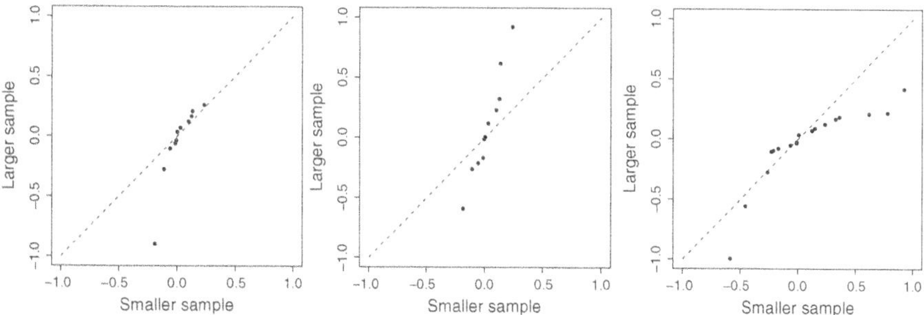

Figure 7.2 Q-Q plots comparing the distributional shapes of the walking directions of: the control group and the first treatment group (left); the control group and the second treatment group (centre); the two treatment groups (right). The origin $(0, 0)$ in each plot corresponds to where the median directions of the two samples coincide

7.2 Tests for a Common Mean Direction

The mean direction is a meaningful measure of central tendency to use with reflectively symmetric unimodal distributions. For such distributions, the mean and median directions coincide.

7.2.1 *Watson's Large-sample Nonparametric Test*

Fisher (1993, Section 5.3.4) describes a large-sample test due to Watson (1983, pages 146–7) for testing the null hypothesis of a common mean direction for two or more distributions. Unlike alternative tests of this null hypothesis, Watson's test does not assume that the underlying distributions have a common dispersion or a common shape.

The test procedure has two variants; the P procedure and the M procedure. In order to decide which one to use, we first need to calculate the values of the circular dispersion (3.16) for each of the g samples, $\hat{\delta}_1, \ldots, \hat{\delta}_g$. If the ratio of the largest of their values divided by the smallest is less than 4, then the P procedure should be used; otherwise the M procedure.

When conducting either variant of the test we next calculate the sample mean directions for the g samples, $\bar{\theta}_1, \ldots, \bar{\theta}_g$. The test statistic for the P variant of the test is

$$Y_g = \frac{2(N - R_P)}{\hat{\delta}_0}, \tag{7.1}$$

where

$$R_P = \sqrt{\hat{C}_P^2 + \hat{S}_P^2}, \quad \hat{C}_P = \sum_{k=1}^{g} n_k \cos\bar{\theta}_k, \quad \hat{S}_P = \sum_{k=1}^{g} n_k \sin\bar{\theta}_k, \quad \hat{\delta}_0 = \sum_{k=1}^{g} \frac{n_k \hat{\delta}_k}{N}.$$

For the M procedure the test statistic is

$$Y_g = 2\left(\sum_{k=1}^{g} \frac{n_k}{\hat{\delta}_k} - R_M\right), \tag{7.2}$$

where

$$R_M = \sqrt{\hat{C}_M^2 + \hat{S}_M^2}, \quad \hat{C}_M = \sum_{k=1}^{g} \frac{n_k \cos\bar{\theta}_k}{\hat{\delta}_k}, \quad \hat{S}_M = \sum_{k=1}^{g} \frac{n_k \sin\bar{\theta}_k}{\hat{\delta}_k}.$$

Watson's large-sample test assumes that all of the g sample sizes are at least 25 (see Section 7.2.2 if they are not). The p-value of the test is established by comparing the observed value of (7.1) or (7.2) with the quantiles of the χ^2_{g-1} distribution.

The function **YgVal** below calculates the appropriate value of (7.1) or (7.2) for the combined sample in the circular data object **cdat** composed of **g** independent samples, the individual sizes of which are contained in **ndat**.

```
YgVal <- function(cdat, ndat, g) {
N <- length(cdat) ; ndatcsum <- cumsum(ndat)
delhat <- 0 ; tbar <- 0
for (k in 1:g) {
sample <- circular(0)
if (k==1) {low <- 0} else
if (k > 1) {low <- ndatcsum[k-1]}
for (j in 1:ndat[k]) { sample[j] <- cdat[j+low] }
tm1 <- trigonometric.moment(sample, p=1)
tm2 <- trigonometric.moment(sample, p=2)
Rbar1 <- tm1$rho; Rbar2 <- tm2$rho ; tbar[k] <- tm1$mu
delhat[k] <- (1-Rbar2)/(2*Rbar1*Rbar1)
}
dhatmax <- max(delhat) ; dhatmin <- min(delhat)
if (dhatmax/dhatmin <= 4) {
CP <- 0 ; SP <- 0 ; dhat0 <- 0
for (k in 1:g) {
CP <- CP+ndat[k]*cos(tbar[k]) ; SP <- SP+ndat[k]*sin(tbar[k])
dhat0 <- dhat0+ndat[k]*delhat[k] }
dhat0 <- dhat0/N
RP <- sqrt(CP*CP+SP*SP) ; Yg <- 2*(N-RP)/dhat0
return(Yg) } else
if (dhatmax/dhatmin > 4) {
CM <- 0 ; SM <- 0 ; Yg <- 0
for (k in 1:g) {
CM <- CM+(ndat[k]*cos(tbar[k])/delhat[k])
SM <- SM+(ndat[k]*sin(tbar[k])/delhat[k])
Yg <- Yg+(ndat[k]/delhat[k]) }
RM <- sqrt(CM*CM+SM*SM) ; Yg <- 2*(Yg-RM)
return(Yg) }
}
```

Two of the sample sizes for the desert ant walking directions are less than 25. Hence, if we apply Watson's large-sample test to them, the p-value obtained must be considered, at best, an approximation to the true p-value. This can be done by running the commands:

```
cdat <- c(cdat1, cdat2, cdat3)
n1 <- length(cdat1) ; n2 <- length(cdat2) ; n3 <- length(cdat3)
ndat <- c(n1, n2, n3) ; g <- 3
YgObs <- YgVal(cdat, ndat, g) ; pchisq(YgObs, g-1, lower.tail=F)
```

which return a value of 0.7522 for Y_g and a p-value of 0.6865. According to the p-value, there is no significant difference between the mean directions of the three underlying distributions.

7.2.2 *Bootstrap Version of Watson's Nonparametric Test*

For situations in which some of the sample sizes are less than 25, as is the case for the ant data, the bootstrap version of Watson's test, due to Fisher and Hall (1991) and described in Fisher (1993, Section 8.4.4), should be used. The details are somewhat involved and to save on space we do not reproduce them here. However, the function **YgTestBoot** available from the website can be used to implement this version of the test. Its extra arguments, in addition to those of the function **YgVal**, are **indsym**, which takes values of 0 (underlying distribution assumed to be asymmetric) and 1 (underlying distribution assumed to be symmetric), and the number of bootstrap samples, **B**.

To illustrate the use of **YgTestBoot**, we apply it to the ant data. From the results reported in Section 7.1.1, it appears reasonable to assume that the underlying distributions for the walking directions of the different groups of ants are reflectively symmetric. Thus, we run the commands:

```
indsym <- 1 ; B <- 9999 ; YgTestBoot(cdat, ndat, g, indsym, B)
```

The estimated p-value obtained when we ran the above code was 0.7101, slightly higher than the p-value of 0.6865 obtained using the large-sample version of the test. Using (1.1), a nominally 95% confidence interval for the true p-value is $(0.7012, 0.7190)$. Hence it would appear that the large-sample version of the test, with the χ^2_{k-1} distribution used as its sampling distribution, is, at least for these data, slightly liberal.

7.2.3 *Watson–Williams Test for von Mises Distributions*

R's **circular** package includes the function **watson.williams.test** to perform an alternative test, introduced by Watson and Williams (1956) and subsequently modified by Stephens (1972), for the homogeneity of mean directions. Unlike the procedures described in the preceding two subsections, the correct usage of the test implemented in **watson.williams.test** requires certain restrictive assumptions to hold. Firstly, the test assumes that any samples are drawn independently from von Mises distributions (see Section 4.3.8). Secondly, it assumes that the underlying von Mises distributions share a common value of the concentration parameter, κ, which is greater than 1.

To illustrate the use of the **watson.williams.test** function we apply it to the walking directions of the two treatment groups of ants. Their concentrations seem to be similar, whilst that for the control group is far higher. For the directions of the two treatment groups, the bias-corrected maximum likelihood estimates of κ, obtained using the **circular** package's **mle.vonmises** function (see Section 6.2.1), are 2.78 and 1.83, respectively. Both are in excess of 1, and the ratio of the largest to the smallest is just 1.52. Assuming the underlying distributions to be von Mises with a common concentration parameter value, running the commands:

```
cdat23 <- c(cdat2, cdat3) ; groupID <- c(rep(1,n2), rep(2,n3))
watson.williams.test(cdat23, group=groupID)
```

returns an F-statistic value of 0.6595 on 1 and 48 degrees of freedom, with an associated p-value of 0.4207. According to this test, then, there is no significant difference between the mean directions of the underlying distributions. This result squares with our earlier findings concerning all three mean directions.

Methods for testing for the equality of the mean directions of two or more von Mises distributions under heteroscedasticity are described in Fisher (1993, Section 5.4.2).

7.3 Tests for a Common Median Direction

The median direction will generally, but not exclusively, be of interest when reflective symmetry has been rejected.

7.3.1 *Fisher's Nonparametric Test*

Fisher (1993, Section 5.3.2) proposed a nonparametric test for a common median direction of two or more distributions. First, the median direction of all N data points in the combined sample, $\tilde{\theta}$, is calculated. For the kth sample, let m_k denote the number of the $\theta_{k1} - \tilde{\theta}, \ldots, \theta_{kn_k} - \tilde{\theta}$ values that are negative, where each $\theta_{kj} - \tilde{\theta}, j = 1, \ldots, n_k$, value is calculated to lie in $(-\pi, \pi]$. We denote the total number of such negative values in the combined sample $\boldsymbol{\theta}$ by $M = m_1 + \cdots + m_g$. The null hypothesis is that the g independent samples were drawn from distributions with a common median direction, and the test statistic is defined as

$$P_g = \frac{N^2}{M(N-M)} \sum_{k=1}^{g} \frac{m_k^2}{n_k} - \frac{NM}{N-M}. \tag{7.3}$$

The null hypothesis is rejected if P_g is judged to be excessively large. When all g sample sizes are at least 10, or for slightly lower sample sizes providing the sample sizes are approximately equal, Fisher (1993, Section 5.3.2) recommends comparison of P_g with the quantiles of the χ^2_{g-1} distribution.

Below, the function **PgVal** makes use of the function **MinusPiPi** to compute the value of P_g for the combined sample in the circular data object **cdat** (consisting of angles measured in radians in $[0, 2\pi)$ or $(-\pi, \pi]$). Its other arguments are the same as those of the function **YgVal**.

```
MinusPiPi <- function(sample) {
n <- length(sample)
for (j in 1:n) {
if (sample[j] < -pi) {sample[j] <- sample[j]+(2*pi)} else
if (sample[j] > pi) {sample[j] <- sample[j]-(2*pi)} }
return(sample)
}

PgVal <- function(cdat, ndat, g) {
N <- length(cdat) ; sumterms <- 0 ; M <- 0
ndatcsum <- cumsum(ndat) ; gmedian <- medianCircular(cdat)
for (k in 1:g) {
if (k==1) {low <- 0} else
if (k > 1) {low <- ndatcsum[k-1]}
sample <- circular(0)
for (j in 1:ndat[k]) { sample[j] <- cdat[j+low] }
shiftdat <- MinusPiPi(sample-gmedian)
m <- length(shiftdat[shiftdat<0]) ; M <- M+m
sumterms <- sumterms + m*m/ndat[k]
}
term1 <- ((N*N)/(M*(N-M))) ; term2 <- (N*M)/(N-M)
Pg <- term1*sumterms-term2 ; return(Pg)
}
```

To illustrate the use of the functions **PgVal** and **MinusPiPi**, we apply them to the data on the walking directions of the three groups of desert ants introduced at the beginning of Section 7.1.1. All three sample sizes are greater than 10. Running the commands:

```
PgObs <- PgVal(cdat, ndat, g) ; pchisq(PgObs, g-1, lower.tail=F)
```

outputs an observed value of 1.6911 for P_g with a p-value of 0.4293. Thus, there would appear to be no significant difference between the median directions of the underlying distributions. Remember, in Section 7.2 we found there was no significant difference between their mean directions.

7.3.2 *Randomization Version of Fisher's Nonparametric Test*

If there is any doubt concerning the appropriateness of using the chi-squared sampling distribution, the significance of the observed value of (7.3) can be established using a randomization version of the test. For each of N_R random permutations of the data values in the original combined sample $\boldsymbol{\theta}$, the first n_1 elements are taken to form the first sample, the next n_2 as the second, and so on. The value of P_g is calculated for the original data set and each one of the N_R random permutations, and the p-value of the test estimated by the proportion of the $(N_R + 1)$ values of P_g that are greater than or equal to that for the original sample.

The function **PgRandTest** below performs the necessary calculations based on **NR** random permutations of the original combined sample.

```
PgRandTest <- function(cdat, ndat, g, NR) {
ndatcsum <- cumsum(ndat)
PgObs <- PgVal(cdat, ndat, g) ; nxtrm <- 1
for (r in 1:NR) {
randsamp <- sample(cdat)
PgRand <- PgVal(randsamp, ndat, g)
if (PgRand >= PgObs) { nxtrm <- nxtrm+1 } }
pval <- nxtrm/(NR+1) ; return (c(PgObs, pval))
}
```

Although the sample sizes for the walking directions of the three groups of ants are all greater than 10, here we apply the randomization version of the test to them in order to illustrate the use of the **PgRandTest** function and compare the p-value returned by it with that obtained in Section 7.3.1. When we ran the commands:

```
NR <- 9999 ; PgRandTest(cdat, ndat, g, NR)
```

the estimated p-value returned was 0.4195. Using (1.1), a nominally 95% confidence interval for the true p-value is $(0.4098, 0.4292)$. Hence, again, there is no evidence of any significant difference between the median directions of the underlying distributions. The p-value calculated using the chi-squared distribution as the sampling distribution (0.4293) lies fractionally above the upper limit of this confidence interval.

7.4 Tests for a Common Concentration

In this testing scenario the null hypothesis of interest specifies that the distributions from which the g samples, $\boldsymbol{\theta}_1, \ldots, \boldsymbol{\theta}_g$, were drawn have a common concentration. This is the circular analogue of homoscedasticity.

7.4.1 *Wallraff's Nonparametric Test*

A simple nonparametric test of circular homoscedasticity is generally credited to Wallraff (1979). To implement it, we first calculate the mean directions of the g samples, $\bar{\theta}_1, \ldots, \bar{\theta}_g$. For the observations in each sample we then calculate the distances (3.19) between them and the mean direction for that sample. A Kruskal–Wallis test is then applied to the distances for the g samples.

The function **WallraffTest** below implements this nonparametric test. Its arguments coincide with those of other functions defined previously in this chapter.

```
WallraffTest <- function(cdat, ndat, g) {
N <- length(cdat) ; ndatcsum <- cumsum(ndat)
tbar <- circular(0) ; distdat <- 0
for (k in 1:g) {
dist <- 0 ; sample <- circular(0)
if (k==1) {low <- 0} else
if (k > 1) {low <- ndatcsum[k-1]}
for (j in 1:ndat[k]) { sample[j] <- cdat[j+low] }
tm1 <- trigonometric.moment(sample, p=1) ; tbar[k] <- tm1$mu
for (j in 1:ndat[k]) { dist[j] <- pi-abs(pi-abs(sample[j]-tbar[k])) }
distdat <- c(distdat, dist) }
distdat <- distdat[-1]
gID <- c(rep(1,n1), rep(2,n2), rep(3,n3))
TestRes <- kruskal.test(distdat, g=gID)
return(TestRes)
}
```

Running it for the walking directions of the three groups of ants, via the command:

```
WallraffTest(cdat, ndat, g)
```

returns a value of 8.5778 for the Kruskal–Wallis chi-squared test statistic on two degrees of freedom and an associated p-value of 0.0137. Thus, the test rejects a common concentration value for a significance level of 1.4% or above.

7.4.2 *Fisher's Test for von Mises Distributions*

Fisher (1993, Section 5.4.4) describes a method he introduced in Fisher (1986) for testing the null hypothesis of a common concentration parameter value for two or more von Mises distributions. Formally, the null hypothesis is H_0: $\kappa_1 = \cdots = \kappa_g = \kappa$, where κ_k, $k = 1, \ldots, g$, denotes the value of the concentration parameter for the distribution from which the kth sample was drawn.

To implement this test, we first calculate the sample mean direction of each sample, so obtaining $\bar{\theta}_1, \ldots, \bar{\theta}_g$. For each sample, the sample mean direction is subtracted from each observation. Thus, for the kth sample, we calculate $(\theta_{k1} - \bar{\theta}_k), \ldots, (\theta_{kn_k} - \bar{\theta}_k)$. The effect of this transformation is to rotate each sample to have a sample mean direction of zero. For the jth observation, $j = 1, \ldots, n_k$, in the kth sample we then calculate the absolute value $d_{kj} = |\sin(\theta_{kj} - \bar{\theta}_k)|$. The mean of the absolute values in the kth sample is given by

$$\bar{d}_k = \sum_{j=1}^{n_k} \frac{d_{kj}}{n_k},$$

and the global mean of the absolute values from all g samples by

$$\bar{d} = \sum_{k=1}^{g} \frac{n_k \bar{d}_k}{N}.$$

Finally, the test statistic F_g is defined as

$$F_g = \frac{(N-g)\sum_{k=1}^{g} n_k(\bar{d}_k - \bar{d})^2}{(g-1)\sum_{k=1}^{g}\sum_{j=1}^{n_k}(d_{kj} - \bar{d}_k)^2}.$$

Below, the function **dValues** returns an object containing all the $d_{kj} = |\sin(\theta_{kj} - \bar{\theta}_k)|$ values for the combined sample in the circular data object **cdat**. The object containing the d_{kj} values is then used as an argument of the **FgVal** function that returns the value of F_g.

```
dValues <- function(cdat, ndat, g) {
N <- length(cdat) ; ndatcsum <- cumsum(ndat) ; dval <- 0
for (k in 1:g) {
sample <- circular(0)
if (k==1) { low <- 0 } else
if (k > 1) { low <- ndatcsum[k-1] }
for (j in 1:ndat[k]) { sample[j] <- cdat[j+low] }
tm1 <- trigonometric.moment(sample, p=1) ; tbar <- tm1$mu
dvalk <- abs(sin(sample-tbar)) ; dval <- c(dval, dvalk) }
dval <- dval[-1] ; return(dval)
}

FgVal <- function(dvals, ndat, g) {
N <- length(dvals) ; ndatcsum <- cumsum(ndat)
sum1 <- 0 ; sum2 <- 0 ; dk <- 0 ; dbar <- 0 ; gdbar <- 0
for (k in 1:g) {
sample <- circular(0)
if (k==1) { low <- 0 } else
if (k > 1) { low <- ndatcsum[k-1] }
for (j in 1:ndat[k]) { dk[j] <- dvals[j+low] }
dbar[k] <- sum(dk)/ndat[k] ; sum2 <- sum2+sum((dk-dbar[k])**2)
gdbar <- gdbar+ndat[k]*dbar[k] }
gdbar <- gdbar/N
for (k in 1:g) { sum1 <- sum1+ndat[k]*(dbar[k]-gdbar)**2 }
Fg <- (N-g)*sum1/((g-1)*sum2) ; return(Fg)
}
```

When all of the sample sizes are at least 10, and the median of the bias-corrected maximum likelihood estimates of the $\kappa_1, \ldots, \kappa_g$ is at least 1, the p-value of the test can be established by comparing the observed value of F_g with the quantiles of the F distribution with $g - 1$ and $N - g$ degrees of freedom.

For the ant data, the bias-corrected maximum likelihood estimates of κ_1, κ_2 and κ_3 are 14.29, 2.778 and 1.827, respectively. The median of the three estimates is $2.778 > 1$, and all three sample sizes are greater than 10. Hence, if we are prepared to assume that the data were drawn from von Mises distributions, we can use the large-sample (and concentrated) version of the test. Running the commands:

```
N <- n1+n2+n3 ; dvals <- dValues(cdat, ndat, g)
FgObs <- FgVal(dvals, ndat, g) ; pf(FgObs, g-1, N-g, lower.tail=F)
```

returns an observed value of 11.6012 for F_g and an associated p-value of 5.7796×10^{-5}. Thus, according to this version of Fisher's test, there is strong evidence of a difference between the concentrations of the distributions from which the three samples were drawn (assuming that they were von Mises). Note, however, that the function **vMGoFBoot** introduced in Section 6.2.3 returns p-values that are all above 0.78 for the directions for the control group; all below 0.01 for those of the first treatment group and all above 0.40 for the directions in the second treatment group. Thus, the assumption of an underlying von Mises distribution for the walking directions of the ants in the first treatment group appears to be unjustified. Formally, then, we should not have applied the test. However, as Fisher (1993, page 132) explains, the F_g-based test is not as sensitive as other tests to the von Mises assumption.

7.4.3 *Randomization Version of Fisher's Test*

If some of the g sample sizes are less than 10, or the median of the bias-corrected maximum likelihood estimates of $\kappa_1, \ldots, \kappa_g$ is less than 1, the p-value of Fisher's test obtained by comparing the observed value of F_g with the quantiles of the F distribution with $g - 1$ and $N - g$ degrees of freedom must be considered, at best, to be an approximation to the true p-value. Instead, the p-value can be estimated using a randomization version of the test in which the d_{kj} values are randomly assigned to the g groups. In fact, this version of the test does not require the underlying distributions to be von Mises; just that they, or there reflections about their mean directions, should share a common shape. Thus, this version of the test provides an alternative to Wallraff's nonparametric test discussed in Section 7.4.1.

The randomization version of the F_g-based test is implemented in the function **FgTestRand** below. Its arguments are the same as those of the function **FgVal**, apart from **NR** which specifies the number of randomized samples to use.

```
FgTestRand <- function(dvals, ndat, g, NR) {
FgObs <- FgVal(dvals, ndat, g) ; nxtrm <- 1
for (r in 1:NR) {
randdvals <- sample(dvals)
FgRand <- FgVal(randdvals, ndat, g)
if (FgRand >= FgObs) { nxtrm <- nxtrm+1 } }
pval <- nxtrm/(NR+1) ; return(pval)
}
```

For the walking directions of the ants, running the commands:

```
NR <- 9999 ; FgTestRand(dvals, ndat, g, NR)
```

returns an estimated p-value of 0.0001, and hence the null hypothesis of underlying distributions with a common concentration value is emphatically rejected. According to this test, the evidence against a common concentration value is even stronger than Wallraff's test suggests.

7.5 Tests for a Common Distribution

In this section we consider procedures for testing whether the data in the g samples, $\boldsymbol{\theta}_1, \ldots, \boldsymbol{\theta}_g$, were drawn from a common distribution.

7.5.1 *Chi-squared Test for Grouped Data*

If the data in each of the g samples are grouped using the same n_c, say, class intervals, or we decide to group the original data into such class intervals, then the data in the g samples can be represented as a $g \times n_c$ contingency table. Testing whether the samples were drawn from a common distribution is then equivalent to testing the contingency table for independence using the standard chi-squared test. For grouped data this will certainly be the simplest approach to adopt. However, when the data are not grouped, the Mardia–Watson–Wheeler and Watson's U^2 tests, considered below, will be more powerful.

7.5.2 *Large-sample Mardia–Watson–Wheeler Test*

This test was first proposed by Wheeler and Watson (1964) for use with two independent samples, and then extended for use with g independent samples by Mardia (1972). The test for the latter scenario proceeds as follows.

The circular data from the g groups are first combined to form the vector $\boldsymbol{\theta}$, and the elements in $\boldsymbol{\theta}$ are then ranked using an arbitrary zero direction. Denoting the rank of the jth element in the kth sample by R_{kj}, the test statistic for the test is then

$$W_g = 2\sum_{k=1}^{g} \frac{C_k^2 + S_k^2}{n_k}, \tag{7.4}$$

where

$$C_k = \sum_{j=1}^{n_k} \cos\left(\frac{2\pi R_{kj}}{N}\right), \quad S_k = \sum_{j=1}^{n_k} \sin\left(\frac{2\pi R_{kj}}{N}\right)$$

and the $2\pi R_{kj}/N$ are referred to as *uniform scores* of the data in $\boldsymbol{\theta}$. This test is essentially a rank-based extension of the Rayleigh test for circular uniformity considered in Section

5.1.1. The null hypothesis of a common distribution is rejected if the observed value of (7.4) is overly large. If all g sample sizes are greater than 10 then the p-value of the test is established by comparing the observed value of W_g with the percentiles of the chi-squared distribution with $2(g-1)$ degrees of freedom.

Below, the function **CosSinUniScores** returns an object containing all the values of $\cos(2\pi R_{kj}/N)$ and $\sin(2\pi R_{kj}/N)$ for the combined sample in the circular data object **cdat**. That object is then used as an argument of the **WgVal** function that computes the corresponding value of W_g.

```
CosSinUniScores <- function(cdat) {
N <- length(cdat)
ranks <- rank(cdat, ties.method="random")
CosUniScores <- cos(ranks*2*pi/N) ; SinUniScores <- sin(ranks*2*pi/N)
return(list(CosUniScores, SinUniScores))
}

WgVal <- function(CSUScores, ndat, g) {
CosUScores <- CSUScores[[1]] ; SinUScores <- CSUScores[[2]]
N <- length(CosUScores) ; ndatcsum <- cumsum(ndat) ; Wg <- 0
for (k in 1:g) {
CosUScoresk <- 0 ; SinUScoresk <- 0
if (k==1) { low <- 0 } else
if (k > 1) { low <- ndatcsum[k-1] }
for (j in 1:ndat[k]) {
CosUScoresk[j] <- CosUScores[j+low] ; SinUScoresk[j] <- SinUScores[j+low] }
sumCkSq <- (sum(CosUScoresk))**2 ; sumSkSq <- (sum(SinUScoresk))**2
Wg <- Wg+(sumCkSq+sumSkSq)/ndat[k] }
Wg <- 2*Wg ; return(Wg)
}
```

Note that, in the function **CosSinUniScores**, ties are broken randomly. Hence, when ties occur the value of W_g obtained will vary between repeated runs of the same code. The effect should be slight unless many ties occur.

For the walking directions of the ants, the sample sizes of the three groups are all greater than 10. When we ran the commands:

```
CSUScores <- CosSinUniScores(cdat)
WgObs <- WgVal(CSUScores, ndat, g) ; pchisq(WgObs, 2*(g-1), lower.tail=F)
```

the value returned for W_g was 7.1014, with an associated p-value of 0.1306. Repeated runs of the same code produced changes in the p-value's second decimal place and beyond. According to this test, then, there is no significant difference between the distributions from which the three samples were drawn. This may seem surprising given the differences in the distributions discussed earlier in the chapter. Such an outcome can be understood in terms of the reduced power that accrues when testing for any difference between distributions rather than a specific type of difference (e.g. between concentrations).

7.5.3 *Randomization Version of the Mardia–Watson–Wheeler Test*

If any of the g sample sizes are less than 10, the p-value obtained using the large-sample version of the Mardia–Watson–Wheeler test discussed above in Section 7.5.2 will be, at best, an approximation to the true p-value. Instead, we can estimate the p-value using a randomization version of the test in which the pairs of $\cos(2\pi R_{kj}/N)$ and $\sin(2\pi R_{kj}/N)$ values are assigned at random to the g groups. This randomization version of the W_g-based test is implemented in the function **WgTestRand** below. Its arguments are the same as those of the function **WgVal**, apart from **NR** which specifies the number of randomizations to use.

```
WgTestRand <- function(CSUScores, ndat, g, NR) {
CosUScores <- CSUScores[[1]] ; SinUScores <- CSUScores[[2]]
N <- length(CosUScores) ; ndatcsum <- cumsum(ndat)
WgObs <- WgVal(CSUScores, ndat, g) ; nxtrm <- 1
ind <- seq(1, N)
for (r in 1:NR) {
CosUScoresRand <- 0 ; SinUScoresRand <- 0
randind <- sample(ind)
for (k in 1:g) {
CosUScoresk <- 0 ; SinUScoresk <- 0
if (k==1) { low <- 0 } else
if (k > 1) { low <- ndatcsum[k-1] }
for (j in 1:ndat[k]) {
CosUScoresk[j] <- CosUScores[randind[j+low]]
SinUScoresk[j] <- SinUScores[randind[j+low]] }
CosUScoresRand <- c(CosUScoresRand, CosUScoresk)
SinUScoresRand <- c(SinUScoresRand, SinUScoresk) }
CosCRanksRand <- CosCRanksRand[-1]
SinUScoresRand <- SinUScoresRand[-1]
CSUScoresRand <- list(CosUScoresRand, SinUScoresRand)
WgRand <- WgVal(CSUScoresRand, ndat, g)
if (WgRand >= WgObs) { nxtrm <- nxtrm+1 } }
pval <- nxtrm/(NR+1) ; return(pval)
}
```

For the walking directions of the three groups of ants, when we ran the commands:

```
NR <- 9999 ; WgTestRand(CSCRanks, ndat, g, NR)
```

the estimated p-value returned was 0.1407, slightly higher than that obtained previously using the large-sample version of the test. Neither, then, does this version of the test provide significant evidence of a difference between the distributions from which the samples were drawn.

7.5.4 *Watson's Two-sample Test*

When $g = 2$, a commonly used alternative to the Mardia–Watson–Wheeler test is Watson's two-sample U^2 test proposed by Watson (1962). This is a particularly attractive option as it is programmed as the **watson.two.test** function within **R**'s **circular** package.

For example, suppose we wanted to test whether the walking directions of the ants in the control group and those for the ants in the second treatment group were drawn from a common distribution. Running the commands:

```
watson.two.test(cdat1, cdat3)
```

returns a test statistic value of 0.1944 and an associated p-value in $(0.01, 0.05)$. Thus, according to this test there is evidence of a significant difference between the two distributions. Of course, if we wanted to use the test for more than this comparison, some allowance should be made for the fact that then we would be performing multiple comparisons. Bonferroni correction is generally the simplest form of correction in such circumstances.

7.5.5 Randomization Version of Watson's Two-sample Test

Critical values for Watson's two-sample U^2 test are available, for example, in Zar (2010, page 849), and these are used by the **watson.two.test** function. Alternatively, a randomization version of the test is always available.

The randomized samples are obtained by randomly permuting the combined sample of size $n_1 + n_2$ formed from the two samples, and using the first n_1 observations as the first sample and the remaining n_2 as the second. For the original data and each of N_r permutations of it, the value of the test statistic is calculated and the p-value of the test estimated by the proportion of the $(N_r + 1)$ values that are greater than or equal to the value of the test statistic for the original sample. As the test statistic is readily calculated using the **circular** package's **watson.two.test** function, the whole process is easy to implement. This has been done in the function **WatsonU2TestRand** below.

```
WatsonU2TestRand <- function(cdat1, cdat2, NR) {
U2Obs <-watson.two.test(cdat1, cdat2)$statistic ; nxtrm <- 1
n1 <- length(cdat1) ; n2 <- length(cdat2) ; N <- n1+n2
combsample <- c(cdat1, cdat2)
for (r in 1:NR) {
randsamp <- sample(combsample)
randsamp1 <- randsamp[1:n1] ; randsamp2 <- randsamp[(n1+1):N]
U2Rand <- watson.two.test(randsamp1, randsamp2)$statistic
if (U2Rand >= U2Obs) { nxtrm <- nxtrm+1 } }
pval <- nxtrm/(NR+1) ; return(c(U2Obs, pval))
}
```

When we ran the randomization version of the test for the walking directions of the control and the second treatment groups of ants, using the commands:

```
NR <- 9999 ; WatsonU2TestRand(cdat1, cdat3, NR)
```

the estimated p-value returned was 0.0386. This p-value lies in the interval quoted at the end of Section 7.5.4. Using (1.1), a nominally 95% confidence interval for the true p-value is $(0.0348, 0.0424)$. So, there is evidence that the distributions from which the two samples were drawn are different at, at least, the 5% significance level.

Note that Maag (1966) suggested a method for extending Watson's U^2 test to more than two samples. Its implementation is slightly more complicated than that for the Mardia–Watson–Wheeler test discussed in Sections 7.5.2 and 7.5.3.

The only other commonly encountered method for testing two samples for a common underlying distribution is that of Kuiper (1960). That test is described fully by Batschelet (1981, Section 6.5) and Upton and Fingleton (1989, Section 9.7). As little work has been published comparing the various procedures, there is no theoretical reason for preferring Watson's U^2 test over any other. We chose to discuss it in detail simply because it has already been programmed in **R**'s **circular** package.

7.6 Moore's Test for Paired Circular Data

In certain situations, two samples of circular data will be paired in the sense that a particular element in one of the samples is linked to the corresponding element of the other sample. Consider, for example, the wind directions at two neighbouring wind farms measured at noon each day. For each day we have one reading for the first sample obviously linked with the corresponding reading in the second. If the distance between the two wind farms were not great, we would expect there to be association between the paired readings in the two samples. A thorough treatment of the analysis of paired circular data is provided by Zar (2010, Chapter 27), a reference we draw on heavily in this section. Here we consider a non-parametric test due to Moore (1980) for testing the null hypothesis that the two samples were drawn from the same distribution.

Generalizing, suppose we have two samples, $\theta_{11}, \ldots, \theta_{1n}$ and $\theta_{21}, \ldots, \theta_{2n}$, where θ_{1j} is paired with θ_{2j}, $j = 1, \ldots, n$. First, each pair of angles $(\theta_{1j}, \theta_{2j})$ is transformed to a pair of rectangular coordinates $(x_j, y_j) \in [-2, 2] \times [-2, 2]$, where $x_j = \cos(\theta_{1j}) - \cos(\theta_{2j})$ and $y_j = \sin(\theta_{1j}) - \sin(\theta_{2j})$. Next, for each $j = 1, \ldots, n$ we calculate

$$r_j = \sqrt{x_j^2 + y_j^2}, \quad \cos\phi_j = \frac{x_j}{r_j}, \quad \sin\phi_j = \frac{y_j}{r_j}.$$

Here, r_j is the magnitude of the vector connecting the origin to (x_j, y_j), and ϕ_j is its direction measured from the positive horizontal axis. The radial distances, the r_j, are then ranked from smallest to largest. Let R_j denote the rank of r_j. The test statistic is

$$R = \sqrt{\frac{\bar{R}_c^2 + \bar{R}_s^2}{n}}, \tag{7.5}$$

where

$$\bar{R}_c = \frac{1}{n}\sum_{j=1}^{n} R_j \cos\phi_j, \quad \bar{R}_s = \frac{1}{n}\sum_{j=1}^{n} R_j \sin\phi_j.$$

Large values of R provide evidence against the null hypothesis of a common distribution for the two samples. Moore (1980) gives critical values for the test, but a randomization version of the test is easily programmed. In the latter, an observation from each original pair is randomly assigned to one of the samples and the other observation from the same original pair assigned to the other sample. Given the form of the test statistic, the only values that possibly change between randomizations are the signs of the $\cos\phi_j$ and $\sin\phi_j$ values. Their values are unchanged if the two observations in a pair are assigned to the same samples as for the original pair. Otherwise, when they are interchanged between the two samples, their values are multiplied by −1. For efficiency, then, below we provide the function **MooreRStats** which returns three objects containing the R_j, $\cos\phi_j$ and $\sin\phi_j$ values for the original paired samples in the *linear* data objects **ldat1** and **ldat2** containing angles measured in radians. The three new objects are then used as arguments of the function **MooreRTestStat** which calculates the value of the test statistic (7.5) and of the function **MooreRTestRand** which performs the randomization version of the test using **NR** randomizations.

```
MooreRStats <- function(ldat1, ldat2) {
x <- cos(ldat1)-cos(ldat2) ; y <- sin(ldat1)-sin(ldat2)
r <- sqrt((x*x)+(y*y)) ; Ranks <- rank(r)
cosphi <- x/r ; sinphi <- y/r
return(list(cosphi, sinphi, Ranks))
}

MooreRTestStat <- function(cosphi, sinphi, Ranks) {
n <- length(cosphi)
RbarC <- (1/n)*sum(Ranks*cosphi) ; RbarS <- (1/n)*sum(Ranks*sinphi)
Rval <- sqrt(((RbarC*RbarC)+(RbarS*RbarS))/n) ; return(Rval)
}

MooreRTestRand <- function(cosphi, sinphi, Ranks, NR) {
RObs <- MooreRTestStat(cosphi, sinphi, Ranks) ; nxtrm <- 1
n <- length(cosphi)
for (r in 1:NR) {
cosphirand <- 0 ; sinphirand <- 0
for (j in 1:n) {
if (runif(1) < 0.5) { cosphirand[j] <- cosphi[j] ; sinphirand[j] <- sinphi[j] }
else { cosphirand[j] <- -cosphi[j] ; sinphirand[j] <- -sinphi[j] } }
RRand <- MooreRTestStat(cosphirand, sinphirand, Ranks)
if (RRand >= RObs) { nxtrm <- nxtrm+1 } }
pval <- nxtrm/(NR+1) ; return(c(RObs, pval))
}
```

To illustrate the use of these three functions, we apply them to data described in Zar (2010, page 655). The data were recorded in an observational study involving 10 birds. In the study, the orientations of the branches on which each bird roosted during the morning and during the afternoon were measured. The pairs of orientations (measured in degrees from north) recorded for the ten birds were: (105, 205), (120, 210), (135, 235), (95, 245), (155, 260), (170, 255), (160, 240), (155, 245), (120, 210), (115, 200). Note that all the afternoon orientations are considerably larger than their morning counterparts. This is also evident from a consideration of the scatterplot in Fig. 7.3, in which all 10 points lie well above the dashed diagonal line. In the raw circular plot, on the right of the same figure, the pairing of the orientations is lost. However, it is clear that all the afternoon orientations are greater than those for the morning. When we ran the commands:

```
ldat1 <- c(105,120,135,95,155,170,160,155,120,115)
ldat2 <-c(205,210,235,245,260,255,240,245,210,200)
ldat1 <- ldat1*2*pi/360 ; ldat2 <- ldat2*2*pi/360
RoostingStats <- MooreRStats(cdat1, cdat2)
cosphi <- RoostingStats[[1]] ; sinphi <- RoostingStats[[2]]
Ranks <- RoostingStats[[3]]
NR <- 9999 ; MooreRTestRand(cosphi, sinphi, Ranks, NR)
```

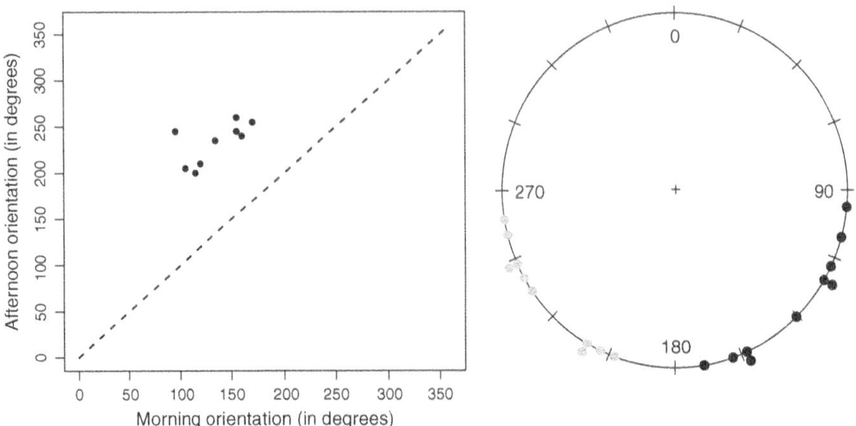

Figure 7.3 Graphical representations of the orientations, measured in degrees from north, of the branches on which 10 birds roosted in the morning and afternoon. Left: scatterplot of the afternoon orientations against the morning orientations with a dashed diagonal line indicating equality of the two orientations. Right: raw circular data plot with the orientations for the morning (afternoon) indicated in black (grey)

the observed value of (7.5) returned was 1.6396, with an estimated p-value of 0.0016. Thus, as we might have expected after a consideration of Fig. 7.3, there is a statistically significant difference between the orientations of the roosting branches used by the birds in the morning and in the afternoon.

8

Correlation and Regression

8.1 Introduction

We introduced the notions of toroidal data, involving joint observations on two circular random variables, and cylindrical data, corresponding to joint observations on one linear random variable and one circular random variable, in Section 1.7. Towards the end of Chapter 4 we briefly discussed potential models for such data. In this chapter we consider correlation coefficients and regression methods that can be used with them. Because they are easier to understand for those used to dealing with linear data, we start by introducing two correlation coefficients used as measures of association for cylindrical data. We also consider tests for independence based on them. We then progress to correlation coefficients for use with toroidal data, and their related tests of independence. The last four sections deal with regression techniques, starting with regression for a linear response and a circular explanatory variable. We move on to consider regression for a circular response and one or more linear regressor variables, and for a circular response and a circular explanatory variable. The chapter ends with a brief discussion of multivariate regression with circular regressors.

8.2 Linear–Circular Association

In this section we consider measures of association for use with cylindrical data, i.e. joint observations on one linear random variable, X, and one circular random variable, Θ. As an illustrative example, we will investigate whether a data set offers support for an association between atmospheric ozone levels at a given locality (a linear variable) and the wind direction at that locality (a circular variable). The various coefficients that have been proposed are generally referred to as measures of *linear–circular association*.

8.2.1 Johnson–Wehrly–Mardia Correlation Coefficient

Mardia (1976) and Johnson and Wehrly (1977) independently proposed a correlation coefficient for use with a random sample, $(x_1, \theta_1), \ldots, (x_n, \theta_n)$, of observations on (X, Θ). To calculate its value we first compute:

$$r_{xc} = r\left((x_1, \ldots, x_n), (\cos\theta_1, \ldots, \cos\theta_n)\right),$$
$$r_{xs} = r\left((x_1, \ldots, x_n), (\sin\theta_1, \ldots, \sin\theta_n)\right),$$
$$r_{cs} = r\left((\cos\theta_1, \ldots, \cos\theta_n), (\sin\theta_1, \ldots, \sin\theta_n)\right),$$

where $r(\mathbf{x}, \mathbf{y})$ denotes the Pearson product moment correlation between the values in the vectors $\mathbf{x}$ and $\mathbf{y}$. The Johnson–Wehrly–Mardia correlation coefficient is then given by

$$R^2_{x\theta} = \frac{r^2_{xc} + r^2_{xs} - 2r_{xc}r_{xs}r_{cs}}{1 - r^2_{cs}}. \quad (8.1)$$

$R^2_{x\theta}$ ranges between zero and one; the greater its value the stronger the association between X and Θ. $R^2_{x\theta}$ is invariant under a change of scale and origin of X as well as under a change of zero or sense of direction for Θ.

The population analogue of $R^2_{x\theta}$ takes the value zero if X is independent of Θ, and the value one if $X = \gamma_0 + \gamma_1 \cos(\Theta - \phi)$, where γ_0, γ_1 and ϕ are constants. If X and Θ are independent, and X is normally distributed, then $(n-3)R^2_{x\theta}/(1 - R^2_{x\theta})$ has an F distribution with 2 and $n - 3$ degrees of freedom. More generally, we recommend testing the null hypothesis of independence using a randomization test in which $R^2_{x\theta}$ is the test statistic.

The function **R2xtCorrCoeff** below calculates the value of $R^2_{x\theta}$ for the *linear* data objects **lvar** and **cvar**, the latter containing angles measured in radians. The function **R2xtIndTestRand** calls **R2xtCorrCoeff** when performing the randomization test of independence based on the use of $R^2_{x\theta}$ as the test statistic.

```
R2xtCorrCoeff <- function(lvar, cvar) {
rxc <- cor(lvar, cos(cvar)) ; rxs <- cor(lvar, sin(cvar))
rcs <- cor(cos(cvar), sin(cvar))
R2xtVal <- ((rxc*rxc)+(rxs*rxs)-(2*rxc*rxs*rcs))/(1-rcs*rcs)
return(R2xtVal)
}

R2xtIndTestRand <- function(lvar, cvar, NR) {
R2xtObs <- R2xtCorrCoeff(lvar, cvar) ; nxtrm <- 1
for (r in 1:NR) {
lvarRand <- sample(lvar)
R2xtRand <- R2xtCorrCoeff(lvarRand,cvar)
if (R2xtRand >= R2xtObs) { nxtrm <- nxtrm+1 } }
pval <- nxtrm/(NR+1) ; return(c(R2xtObs, pval))
}
```

To illustrate the randomization test, we apply it to data on ozone concentration and wind direction (in degrees) published by Johnson and Wehrly (1977) and reproduced in Table 11.1 of Mardia and Jupp (1999). The observed vectors, (x_j, θ_j), $j = 1, \ldots, 19$, are: (28.0, 327), (85.2, 91), (80.5, 88), (4.7, 305), (45.9, 344), (12.7, 270), (72.5, 67),

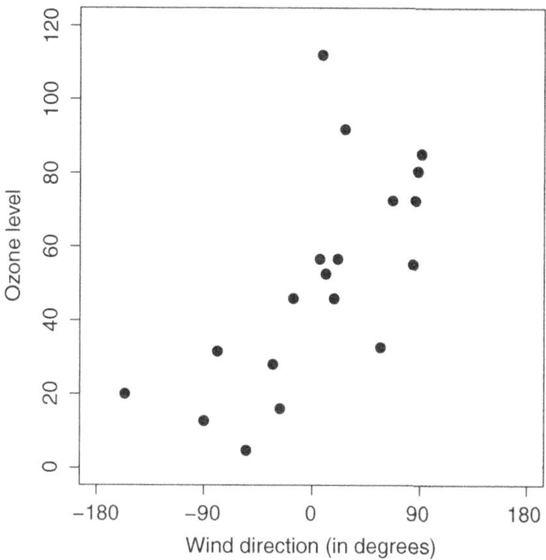

Figure 8.1 Scatterplot of ozone concentration against wind direction with the latter represented as an angle in $(-180^\circ, 180^\circ)$

$(56.6, 21)$, $(31.5, 281)$, $(112.0, 8)$, $(20.0, 204)$, $(72.5, 86)$, $(16.0, 333)$, $(45.9, 18)$, $(32.6, 57)$, $(56.6, 6)$, $(52.6, 11)$, $(91.8, 27)$ and $(55.2, 84)$. A scatterplot of ozone concentration against wind direction represented as an angle in $(-180^\circ, 180^\circ)$ is presented in Fig. 8.1. From that diagram, it appears that the lower ozone concentrations are associated with wind directions between -180° and 0° (or between 180° and 360° for the original data), whereas higher ozone concentrations are associated with wind directions between 0° and 180°. Running the commands:

```
ozone <- c(28.0,85.2,80.5,4.7,45.9,12.7,72.5,56.6,31.5,112.0,20.0,72.5,16.0,45.9,32.6,56.6,52.6,91.8,
55.2)
winddeg <- c(327,91,88,305,344,270,67,21,281,8,204,86,333,18,57,6,11,27,84)
windrad <- winddeg*2*pi/360
R2xtIndTestRand(ozone, windrad, 9999)
```

returns an observed value of $R^2_{x\theta}$ of 0.5220 and, when we ran the code, an estimated p-value for the test for independence of 0.0027. With such a small p-value, independence is emphatically rejected. Hence we conclude that there does appear to be some linear–circular relation between ozone level and wind direction. However, it is important to note that this approach to measuring linear–circular association rests on the assumption that the association is approximately sinusoidal. That is, the data in a scatterplot with the linear variable on the vertical axis and the circular variable on the horizontal axis should be distributed around a sine wave. Figure 8.1 is not strongly supportive of such a pattern.

8.2.2 Mardia's Rank Correlation Coefficient

Following the logic of Spearman's rank correlation coefficient, Mardia (1976) proposed a rank-based analogue of $R^2_{x\theta}$ in which the x_j and θ_j are replaced by their ranks and uniform scores, respectively. This measure of linear–circular association makes the less restrictive assumption that there is only one maximum and one minimum in the relationship between the two variables.

To calculate the coefficient, the original data vectors, $(x_1, \theta_1), \ldots, (x_n, \theta_n)$, are first re-ordered so that $x_1 \leq \cdots \leq x_n$. Next, the uniform scores of their associated angles, the θ_j, are calculated. Thus, if r_j is the linear rank (relative to an arbitrary origin) of the angle θ_j paired with x_j, then its uniform score is $2\pi r_j/n$. The x_j are then replaced by their ranks. The value of $nR^2_{x\theta}$ calculated for $(1, 2\pi r_1/n), \ldots, (n, 2\pi r_n/n)$ is the value of Mardia's rank correlation coefficient,

$$U_n = \frac{24(T_c^2 + T_s^2)}{n^2(n+1)}, \tag{8.2}$$

where

$$T_c = \sum_{j=1}^{n} j \cos\left(\frac{2\pi r_j}{n}\right), \quad T_s = \sum_{j=1}^{n} j \sin\left(\frac{2\pi r_j}{n}\right).$$

A test of independence based on U_n rejects independence if the observed value of U_n is large in comparison with the percentiles of its sampling distribution under independence. For large n, the sampling distribution of U_n under independence is approximately chi-squared on two degrees of freedom if X and Θ have continuous distributions. Appendix 2.17 of Mardia and Jupp (1999) provides quantiles of U_n, under the same assumptions, for various values of n. The test is invariant under a change of origin for the linear variable or under rotation of the circular one.

Alternatively, the significance of $U^* = (T_c^2 + T_s^2)$ can be established using a randomization version of the test in which the uniform scores are randomly assigned to the ranks of the linear variable. This approach is implemented in an efficient way using the three functions below. The uniform scores for the original data objects **lvar** and **cvar** are returned by the function **UniformScores**. The function **Ustar** calculates the value of U^* for the supplied uniform scores, and is called by the function **MardiaRankIndTestRand** which performs the randomization test.

```
UniformScores <- function(lvar, cvar) {
ranklvar <- rank(lvar, ties.method="random")
n <- length(cvar) ; cvar2 <- 0
for (j in 1:n) { cvar2[ranklvar[j]] <- cvar[j] }
rankcvar <- rank(cvar2, ties.method="random")
uscores <- rankcvar*2*pi/n ; return(uscores)
}

Ustar <- function(uniscores) {
n <- length(uniscores) ; Tc <- 0 ; Ts <- 0
```

```
for (j in 1:n) {
Tc <- Tc+j*cos(uniscores[j]) ; Ts <- Ts+j*sin(uniscores[j]) }
Ustar <- (Ts*Ts)+(Tc*Tc) ; return(Ustar)
}

MardiaRankIndTestRand <- function(uniscores, NR) {
UstarObs <- Ustar(uniscores) ; nxtrm <- 1
for (r in 1:NR) {
uniscoresRand <- sample(uniscores) ; UstarRand <- Ustar(uniscoresRand)
if (UstarRand >= UstarObs) { nxtrm <- nxtrm+1 } }
pval <- nxtrm/(NR+1) ; return(c(UstarObs, pval))
}
```

For the ozone concentration data, when we ran the commands:

```
uniscores <- UniformScores(ozone, windrad)
MardiaRankIndTestRand(uniscores, 9999)
```

the estimated p-value returned was 0.0271, indicating, once more, association between the two variables. In the function **UniformScores**, ties are broken randomly, and we note that three of the ozone concentrations are each repeated twice. Such ties will have an effect on the p-value obtained in different runs of the same code, over and above any effect due to randomization. However, the effect should be slight unless many ties occur. The p-value returned here is around 10 times bigger than its counterpart obtained in Section 8.2 using the original data rather than their ranks. This is to be expected because of the rank-based test's less restrictive assumptions.

Another approach to measuring linear–circular association was suggested by Fisher and Lee (1981) and developed further in Fisher (1993, Section 6.2.2). We do not elaborate on it here because it is computationally and conceptually more involved than the approach described above. Interested readers should consult the two references above as well as Mardia and Jupp (1999, Section 11.2.1).

8.3 Circular–Circular Association

In this section we consider measures of association for use with toroidal data, i.e joint observations on two circular random variables. Numerous correlation coefficients for so-called *circular–circular association* have been proposed in the literature, a review of them being provided by Jupp and Mardia (1989). Note, however, that relatively little work has focussed on comparing the performance of the different coefficients.

8.3.1 Fisher–Lee Correlation Coefficient for Rotational Dependence

The Fisher–Lee circular–circular correlation coefficient is a measure of *rotational dependence* of the form

$$\Theta = \Psi + \xi \pmod{2\pi} \quad \text{or} \quad \Theta = -\Psi + \xi \pmod{2\pi},$$

where ξ is a constant angle. Identifying by (Θ_1, Ψ_1) and (Θ_2, Ψ_2) two independent random vectors with the same distribution as (Θ, Ψ), Fisher and Lee (1983) proposed the correlation coefficient

$$\rho_{FL} = \frac{E[\sin(\Theta_1 - \Theta_2)\sin(\Psi_1 - \Psi_2)]}{\{E[\sin^2(\Theta_1 - \Theta_2)]E[\sin^2(\Psi_1 - \Psi_2)]\}^{1/2}}. \tag{8.3}$$

This measure of rotational dependence has much in common with Pearson's product moment coefficient, the standard measure of linear dependence, and has the following properties:

(i) $-1 \leq \rho_{FL} \leq 1$;
(ii) $\rho_{FL} = 1$ if and only if $\Theta = \Psi + \xi \pmod{2\pi}$;
(iii) $\rho_{FL} = -1$ if and only if $\Theta = -\Psi + \xi \pmod{2\pi}$;
(iv) ρ_{FL} is invariant to changes in origin of Θ and Ψ;
(v) reflection of one of Θ or Ψ changes the sign but not the magnitude of ρ_{FL};
(vi) if Θ and Ψ are independent then $\rho_{FL} = 0$;
(vii) if Θ and Ψ are unimodal and highly concentrated then ρ_{FL} is well approximated by Pearson's product moment correlation coefficient.

Given a random sample of n observations of (Θ, Ψ), $(\theta_1, \psi_1), \ldots, (\theta_n, \psi_n)$, the natural estimate of ρ_{FL}, r_{FL}, can be calculated as

$$r_{FL} = \frac{4(AB - CD)}{\{(n^2 - E^2 - F^2)(n^2 - G^2 - H^2)\}^{1/2}},$$

where

$$A = \sum_{j=1}^{n} \cos\theta_j \cos\psi_j, \quad B = \sum_{j=1}^{n} \sin\theta_j \sin\psi_j,$$

$$C = \sum_{j=1}^{n} \cos\theta_j \sin\psi_j, \quad D = \sum_{j=1}^{n} \sin\theta_j \cos\psi_j,$$

$$E = \sum_{j=1}^{n} \cos 2\theta_j, \quad F = \sum_{j=1}^{n} \sin 2\theta_j, \quad G = \sum_{j=1}^{n} \cos 2\psi_j, \quad H = \sum_{j=1}^{n} \sin 2\psi_j.$$

For small-sized samples, with $n < 25$, a randomization test can be used to establish whether r_{FL} is significantly different from 0. To demonstrate this approach we consider a data set taken from Upton and Fingleton (1989, p. 298) of the directions, measured in degrees, travelled by two sea stars (a type of marine animal characterized by having five arms) for each of 10 successive days. The two sea stars began the study just a few centimetres apart.

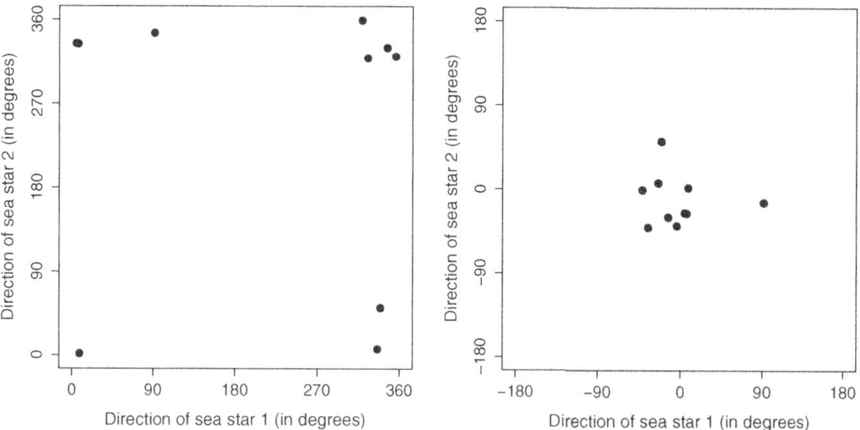

Figure 8.2 Planar representations of the directions travelled by two sea stars on each of 10 successive days, with the unit torus sliced at $(0°, 0°)$ (left) and $(-180°, -180°)$ (right)

A positive correlation between their directions might suggest that both responded similarly to some shared environmental factor each day, such as the prevailing state of the tide.

Two graphical representations of the data are provided in Fig. 8.2. Remember, being bivariate circular, the data really live on the unit torus. The plot on the left can be thought to have been obtained by first slicing the unit torus vertically at the origin for the direction of sea star 1 to form a cylinder. Slicing that cylinder horizontally at the origin for the direction of sea star 2 and flattening the resulting surface produces the planar representation of the data. To recover the original torus, the horizontal lines corresponding to 0° and 360° on the vertical axis must be folded over and glued together to form a unit cylinder, and then the two ends of that cylinder brought together and glued to form the unit torus. In the planar plot, the four points with coordinates $(0°, 0°)$, $(0°, 360°)$, $(360°, 0°)$ and $(360°, 360°)$ all correspond to the origin on the torus. This is not a good representation of the data which are clustered around the origin. Slicing the torus at $(-180°, -180°)$, rather than at $(0°, 0°)$, results in the plot on the right of Fig. 8.2. This representation of the data is far easier to interpret and suggests that there is little or no relation between the two sets of directions.

Below, the function **rFLCorrCoeff** computes the value of r_{FL} for the two *linear* data objects **lcdat1** and **lcdat2** containing angles measured in radians. The function **rFLIndTestRand** performs the randomization test of independence for the same two data objects using **NR** random pairings of their observations.

```
rFLCorrCoeff <- function(lcdat1, lcdat2) {
A <- sum(cos(lcdat1)*cos(lcdat2)) ; B <- sum(sin(lcdat1)*sin(lcdat2))
C <- sum(cos(lcdat1)*sin(lcdat2)) ; D <- sum(sin(lcdat1)*cos(lcdat2))
E <- sum(cos(2*lcdat1)) ; F <- sum(sin(2*lcdat1))
G <- sum(cos(2*lcdat2)) ; H <- sum(sin(2*lcdat2)) ; n <- length(lcdat1)
denom <- sqrt(((n*n)-(E*E)-(F*F))*((n*n)-(G*G)-(H*H)))
rFLVal <- 4*((A*B)-(C*D))/denom ; return(rFLVal)
}
```

```
rFLIndTestRand <- function(lcdat1, lcdat2, NR) {
rFLObs <- rFLCorrCoeff(lcdat1, var2) ; nxtrm <- 1
for (r in 1:NR) {
lcdat1Rand <- sample(lcdat1) ; rFLRand <- rFLCorrCoeff(lcdat1Rand, lcdat2)
if (abs(rFLRand) >= abs(rFLObs)) { nxtrm <- nxtrm+1 } }
pval <- nxtrm/(NR+1) ; return(c(rFLObs, pval))
}
```

For the sea star data, running the commands:

```
star1deg <- c(335.8, 91.2, 339.1, 318.8, 8.6, 7.0, 355.9, 4.6, 346.5, 325.0)
star2deg <- c(6.8, 345.5, 51.2, 359.3, 1.7, 333.9, 320.6, 334.4, 329.9, 318.8)
star1rad <- star1deg*2*pi/360 ; star2rad <- star2deg*2*pi/360
rFLIndTestRand (star1rad, star2rad, 9999)
```

returns a value of –0.1398 for r_{FL} and, when we ran the code, an estimated p-value of 0.6637. Using (1.1), a nominally 95% confidence interval for the true p-value of the test is $(0.6544, 0.6730)$. Thus, there is no evidence of any significant rotational dependence between the daily directions taken by the two sea stars.

For larger sample sizes, with $n \geq 25$, the asymptotic theory described in Fisher (1993, Section 6.3.3) can be appealed to. For such sample sizes, the results obtained using that approach and the randomization approach implemented above should not differ greatly. With ever-increasing personal computing power at ever-reduced cost, the randomization approach can be used to check the results obtained using asymptotic theory even for relatively large sample sizes. Should they differ substantially, the results obtained using the randomization approach should be considered the more reliable.

A nominally $100(1-\alpha)\%$ confidence interval for ρ_{FL} can be obtained using a bootstrap approach in which pairs of the original observations are sampled with replacement. Such an approach is implemented in the function **rhoFLCIBoot.**

```
rhoFLCIBoot <- function(lcdat1, lcdat2, ConfLevel, B) {
alpha <- (100-ConfLevel)/100 ; n <- length(lcdat1) ; rFL <- 0
for (b in 1:B) {
randind <- sample(1:n, n, replace=T)
boot1 <- lcdat1[randind] ; boot2 <- lcdat2[randind]
rFL[b] <- rFLCorrCoeff(boot1, boot2) }
rFL[B+1] <- rFLCorrCoeff(lcdat1, lcdat2) ; rFLsort <- sort(rFL)
return (c(rFLsort[(alpha/2)*(B+1)], rFLsort[(1-alpha/2)*(B+1)]))
}
```

For the sea star data, when we ran the command:

```
rhoFLCIBoot(star1rad, star2rad, 95, 9999)
```

the nominally 95% confidence interval for ρ_{FL} returned was $(-0.6166, 0.4409)$. This interval includes the value zero and so confirms that we have no significant evidence supporting rotational dependence.

For the same data, Upton and Fingleton (1989, page 304) used a jack-knife procedure to calculate a nominally 95% confidence interval for ρ_{FL}, obtaining $(-0.196, -0.090)$. However, their method incorporated a large-sample normal approximation. The results from our bootstrapping procedure suggest that the approximation they used can lead to unreliable estimates for small-sized samples. This likely explains why Fisher (1993, Section 6.3.3) only recommends the jack-knife method when $n > 25$. The jack-knife method is computationally more efficient for such sample sizes but, with the availability of greater and cheaper computing power, the bootstrap method implemented above will be practical for all but vast data sets.

8.3.2 *Fisher–Lee Correlation Coefficient for Toroidal-Monotonic Association*

As explained previously, the coefficient r_{FL} introduced in the preceding subsection is a measure of *rotational* dependence. Fisher and Lee (1983) explain how a small modification in its definition produces a measure of more general *toroidal-monotonic* association. This is achieved by simply replacing the original data by their uniform scores. Thus, if $r_1, \ldots, r_n$ and $s_1, \ldots, s_n$ are the standard linear ranks of $\theta_1, \ldots, \theta_n$ and $\psi_1, \ldots, \psi_n$ relative to an arbitrary common origin, respectively, then their uniform scores are simply the linear ranks multiplied by $2\pi/n$. The resulting measure is analogous to Spearman's rank correlation coefficient for linear data, the latter being the value of Pearson's product moment correlation coefficient obtained by replacing the original data by their ranks.

The r_{FL} coefficient and its uniform scores based counterpart are measures of rotational and toroidal-monotonic associations, respectively. Low absolute values of them cannot, therefore, on there own, be taken as evidence of no association, since a strong but non-monotonic association can produce low absolute values of both measures. Thus, their values should be interpreted in tandem with a plot of the circular data concerned. The two panels of Fig. 8.2 highlight the importance of representing the data suitably so as to meaningfully interpret any relations that might exist between two circular variables.

8.3.3 *Jammalamadaka–Sarma Correlation Coefficient*

The only circular–circular correlation coefficient that has been programmed in **R**'s **circular** library is a circular version of Pearson's product moment correlation proposed by Jammalamadaka and Sarma (1988) and described in Jammalamadaka and SenGupta (2001, Section 8.3). The **cor.circular** function returns: the observed value of their correlation coefficient; the value of a test statistic for testing the null hypothesis that the population correlation coefficient is 0; and the p-value of the test calculated using a large-sample normal approximation. By analogy with Pearson's product moment correlation, if the two circular random variables are independent then the population value of the Jammalamadaka–Sarma correlation coefficient is 0; but a value of 0 for it does not necessarily imply that the two random variables are independent.

To illustrate the use of the **cor.circular** function we apply it to the data on the morning and afternoon roosting orientations of 10 birds introduced in Section 7.6. Running the commands:

```
lmorn <- c(105,120,135,95,155,170,160,155,120,115)
laft <- c(205,210,235,245,260,255,240,245,210,200)
cmorn <- circular(lmorn*2*pi/360) ; caft <- circular(laft*2*pi/360)
JSCorrRes <- cor.circular(cmorn, caft, test=T)
```

returns a value of the correlation coefficient of 0.6716, a test statistic value of 2.1030 and a p-value of 0.0355 for the test of the null hypothesis that the population correlation coefficient is 0. According to these results, then, the null hypothesis can be rejected at around the 4% significance level or above.

As the relation between the performance of the asymptotic normal approximation and sample size has not been explored, for small samples we recommend instead a randomization test for independence, the test statistic of which is the Jammalamadaka–Sarma correlation coefficient. The function **JSTestRand** below implements this test for the two *circular* data objects **cdat1** and **cdat2**, and **NR** randomizations of them.

```
JSTestRand <- function(cdat1, cdat2, NR) {
CorrJSObs <- cor.circular(cdat1, cdat2) ; nxtrm <- 1
for (r in 1:NR) {
cdat1Rand <- sample(cdat1) ; CorrJSRand <- cor.circular(cdat1Rand, cdat2)
if (abs(CorrJSRand) >= abs(CorrJSObs)) { nxtrm <- nxtrm+1 } }
pval <- nxtrm/(NR+1) ; return(c(CorrJSObs, pval))
}
```

When we ran the command:

```
JSTestRand(cmorn, caft, 9999)
```

for the bird orientation data, the estimated p-value returned was 0.0392. Using (1.1), a nominally 95% confidence interval for the true p-value is (0.0354, 0.0430). Note that the p-value of the test based on the asymptotic normal approximation lies within this interval. According to the values included in the confidence interval, independence can be rejected at, at least, the 5% level.

8.3.4 *Rothman's Test for Independence*

We finish this section with a very general test of the null hypothesis that Θ and Ψ are independent proposed by Rothman (1971). Following the description of the test procedure provided by Mardia and Jupp (1999, page 253), we first calculate the standard linear ranks of $\theta_1, \ldots, \theta_n$ and $\psi_1, \ldots, \psi_n$ using any arbitrary common origin, $r_1, \ldots, r_n$ and $s_1, \ldots, s_n$. The test statistic is then

$$A_n = \frac{1}{n^4} \sum_{j=1}^{n} \sum_{k=1}^{n} (T_{jj} + T_{kk} - T_{jk} - T_{kj})^2, \tag{8.4}$$

where

$$T_{jk} = n \min(r_j, s_k) - r_j s_k$$

for $j, k = 1, \ldots, n$, with $\min(x, y)$ denoting the minimum value of x and y. The null hypothesis is rejected if the observed value of A_n is large in comparison with the percentiles of its sampling distribution under independence. For large n and x, Jupp and Spurr (1985) show that $P(16\pi^4 A_n > x) \simeq (1.466x - 0.322)e^{-x/2}$ and give large-sample critical values for $16\pi^4 A_n$ that transform to critical values for A_n of 0.0064, 0.0075 and 0.0099 for the 10%, 5% and 1% significance levels, respectively. However, no results have been published regarding the sample sizes for which these critical values can be applied.

Instead, for small to medium sample sizes, the significance of the observed value of A_n can be established using a randomization version of the test, with the randomized samples consisting of random permutations of the ranks in the two samples. To increase efficiency, this version of the test is implemented below using the three functions: **Ranks**, which calculates the ranks for the two original samples; **RothmanAn**, which uses the ranks as arguments and calculates the value of A_n and **RothmanAnTestRand**, which calls the other two functions when performing the randomization version of the test.

```
Ranks <- function(lcdat1, lcdat2) {
rank1 <- rank(lcdat1, ties.method="random")
rank2 <- rank(lcdat2, ties.method="random")
return(list(rank1, rank2))
}

RothmanAn <- function(rank1, rank2) {
n <- length(rank1) ; AnVal <- 0
for (j in 1:n) {
for (k in 1:n) {
Tjj <- n*min(rank1[j],rank2[j])-(rank1[j]*rank2[j])
Tkk <- n*min(rank1[k],rank2[k])-(rank1[k]*rank2[k])
Tjk <- n*min(rank1[j],rank2[k])-(rank1[j]*rank2[k])
Tkj <- n*min(rank1[k],rank2[j])-(rank1[k]*rank2[j])
AnVal <- AnVal+(Tjj+Tkk-Tjk-Tkj)**2 } }
AnVal <- AnVal/(n**4) ; return(AnVal)
}

RothmanAnTestRand <- function (rank1, rank2, NR) {
AnObs <- RothmanAn(rank1, rank2) ; nxtrm <- 1
for (r in 1:NR) {
randrank1 <- sample(rank1)
AnRand <- RothmanAn(randrank1, rank2)
if (AnRand >= AnObs) { nxtrm <- nxtrm+1 } }
pval <- nxtrm/(NR+1) ; return(c(AnObs, pval))
}
```

For the sea star data, running the commands:

```
SeaStarRanks <- Ranks(star1rad, star2rad)
rank1 <- SeaStarRanks[[1]] ; rank2 <- SeaStarRanks[[2]]
RothmanAnTestRand(rank1, rank2, 9999)
```

returns an observed value of A_n of 0.7826 and, when we ran the code, an estimated p-value for the test of 0.2176. Using (1.1), a nominally 95% confidence interval for the true p-value of the test is (0.2095, 0.2257). Thus, according to this test, there is no significant evidence of an association between the movements of the two sea stars. Note that in the function **Ranks** ties are, again, broken at random. For the sea star data there are no repeated values but for samples that do contain ties the p-value will be affected by them.

8.4 Regression for a Linear Response and a Circular Regressor

In many situations involving cylindrical data we will want to model the relation between a linear response variable and a circular explanatory variable. This is relatively common, for instance, when investigating annual or daily patterns. As an example, Hand *et al.* (1994) provide the data represented in Fig. 8.3 on the number of monthly deaths attributed to lung disease in a given area, collected over six consecutive years. The scatterplot suggests that the seasonal relationship between the two variables is close to sinusoidal. A value of 0.8594 for the Johnson–Wehrly–Mardia correlation coefficient (8.1) supports this. The modelling of data such as these can, happily, often be tackled using standard regression techniques.

8.4.1 Basic Cosine Regression Model

A simple potential model capable of describing the main features of data of the type described above is

$$X_t = \gamma_0 + \gamma_1 \cos(\omega t - \phi) + \epsilon_t, \tag{8.5}$$

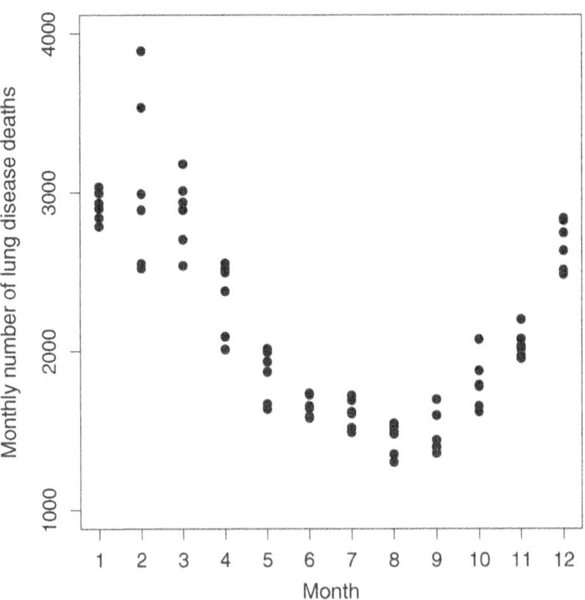

Figure 8.3 Scatterplot of monthly numbers of lung disease deaths for the six-year period of 1974–79

where: X_t denotes the value of the linear response variable at time t; γ_0 is the mean value of the response; γ_1 is the amplitude of the cyclic fluctuation in the response; ω is the angular frequency; ϕ is the so-called acrophase angle and, here and for all subsequent models in this chapter, the ϵ_t are assumed to be independent and identically distributed errors from a normal distribution with mean 0 and constant variance σ^2. From an inspection of Fig. 8.3 there is perhaps evidence that the homoscedasticity assumption might be questionable for the monthly lung disease deaths. We can usually identify ω from our knowledge of the system. In the case of the lung disease deaths, where the 'time' variable t corresponds to the month of the year and takes values $1, \ldots, 12$, the period (T) is 12 and ω is $2\pi/T = \pi/6$. The parameters γ_0, γ_1 and ϕ are unknown constants which must be estimated. Note that in the case where ω cannot be defined from knowledge of the system, the problem becomes considerably more challenging and comes within the scope of time-series analysis rather than the regression methodology considered here.

In practice, rather than fit (8.5), we fit the equivalent model

$$X_t = \gamma_0 + \gamma_1^* \cos(\omega t) + \gamma_2^* \sin(\omega t) + \epsilon_t, \tag{8.6}$$

where $\gamma_1 = \sqrt{\gamma_1^{*2} + \gamma_2^{*2}}$ and $\phi = \text{atan2}(\gamma_2^*, \gamma_1^*)$, with atan2 as defined in (3.5). In **R** we can fit (8.6) using the **lm** function for fitting general linear models. For the monthly lung disease deaths, running the code:

```
lung74 <- c(3035,2552,2704,2554,2014,1655,1721,1524,1596,2074,2199,2512)
lung75 <- c(2933,2889,2938,2497,1870,1726,1607,1545,1396,1787,2076,2837)
lung76 <- c(2787,3891,3179,2011,1636,1580,1489,1300,1356,1653,2013,2823)
lung77 <- c(2996,2523,2540,2520,1994,1641,1691,1479,1696,1877,2032,2484)
lung78 <- c(2899,2990,2890,2379,1933,1734,1617,1495,1440,1777,1970,2745)
lung79 <- c(2841,3535,3010,2091,1667,1589,1518,1349,1392,1619,1954,2633)
alllung <- c(lung74,lung75,lung76,lung77,lung78,lung79)
month <- rep(seq(1,12), 6) ; omega <- pi/6
cosmonth <- cos(omega*month) ; sinmonth <- sin(omega*month)
lung.lmod <- lm(alllung ~ cosmonth+sinmonth)
plot(month, alllung, xlab="Month", ylab="Monthly number of lung disease deaths", ylim=
c(1000,4000), xlim=c(1,12), xaxp=c(0,12,12), yaxp=c(1000,4000,3), cex.axis=1.2, cex.lab=1.4, pch=
16, cex=1.2, lines(predict(lung.lmod), lwd=2))
```

fits (8.6) to the data and produces the left-hand scatterplot in Fig. 8.4 of the data with straight lines connecting the fitted values. Visually, the model appears to capture the main features of the relationship between the variables reasonably well but there this also some evidence of lack of fit.

A more formal diagnosis can be based on the fitted values and the studentized residuals. Both are easily extracted using **R**'s **MASS** package via the commands:

```
install.packages(MASS) ; library(MASS)
lung.pred <- fitted(lung.lmod) ; lung.sres <- studres(lung.lmod)
```

We can produce a scatterplot of the studentized residuals against the fitted values, run the Shapiro–Wilk test for normality, and the Bartlett and Fligner–Killeen tests for the homogeneity of variance, using the commands:

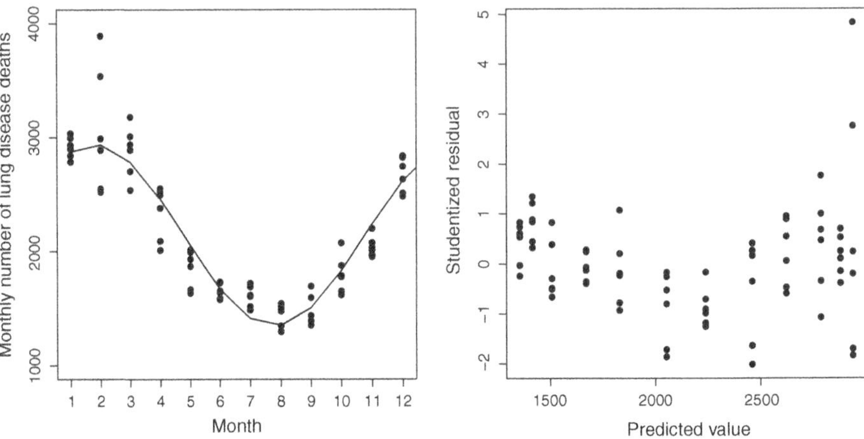

Figure 8.4 Scatterplots of: left, the monthly numbers of lung disease deaths with straight lines connecting the fitted values for model (8.6); right, the studentized residuals against the fitted values for the same fit

```
plot(lung.pred, lung.sres, xlab="Predicted value", ylab="Studentized residual", cex.axis=1.2,
cex.lab=1.4, pch=16, cex=1.2)
shapiro.test(lung.sres)
bartlett.test(lung.sres, month) ; fligner.test(lung.sres, month)
```

The scatterplot produced appears on the right of Fig. 8.4, and the p-values returned for the three tests are 7.657×10^{-5}, 1.337×10^{-6} and 0.0699, respectively. Thus normality is emphatically rejected. Homoscedasticity is rejected by Bartlett's test, but not by the nonparametric Fligner–Killeen test. Bartlett's test is known to be very sensitive to the normality assumption. Further investigation of the performance of the three tests found they depended very heavily on the two studentized residuals with values of around 3 and 5, respectively, clearly visible in Fig. 8.4. Those residuals correspond to two very large monthly totals of 3891 and 3535 deaths for the month of February in 1976 and 1979, respectively. If they are replaced by missing values (NA), and the model refitted, neither assumption is rejected by the tests. There is evidence, therefore, that the numbers of lung disease deaths in February are more variable than those for the other months. Moreover, it is clear from the systematic pattern (decreasing trend followed by an increasing trend) manifested in the right-hand scatterplot of Fig. 8.4 that unexplained structure remains after fitting the cosine regression model.

8.4.2 *Extended Cosine Regression Model*

The model (8.5) can easily to be extended to include more cosine terms, the general equation for such an extended model being

$$X_t = \gamma_0 + \gamma_1 \cos(\omega t - \phi_1) + \gamma_2 \cos(2\omega t - \phi_2) + \cdots + \gamma_k \cos(k\omega t - \phi_k) + \epsilon_t. \qquad (8.7)$$

For illustrative purposes, here we consider fitting the two-component special case of (8.7) to the monthly lung disease deaths data with the observations for February 1976 and 1979 replaced by missing values. Their omission should not affect the fitted values greatly. A more rigorous analysis, involving all the data, would be based on a more complex model capable of modelling the greater variability in the February deaths. The form of the model we actually fit is

$$X_t = \gamma_0 + \gamma_1^* \cos(\omega t) + \gamma_2^* \sin(\omega t) + \gamma_3^* \cos(2\omega t) + \gamma_4^* \sin(2\omega t) + \epsilon_t. \quad (8.8)$$

The fitting can be done using the commands:

```
is.na(lung76) <- c(2) ; is.na(lung79) <- c(2)
alllung <- c(lung74,lung75,lung76,lung77,lung78,lung79)
cosmonth <- cos(omega*month) ; sinmonth <- sin(omega*month)
cos2month <- cos(2*omega*month) ; sin2month <- sin(2*omega*month)
lung2.lmod <- lm(alllung~cosmonth+sinmonth+cos2month+sin2month)
```

The resulting fit is a big improvement on that for (8.5). The scatterplot of the residuals against the fitted values displays no obvious systematic pattern, and the Shapiro–Wilk and Bartlett tests, with p-values of 0.6814 and 0.0886, respectively, provide no significant evidence against the normality and homoscedasticity assumptions. However, the summary information for the fit of the model, returned on using the **summary(lung2.lmod)** command, indicates that the $\cos(2\omega t)$ term is superfluous (p-value of 0.8904 in a t-test of the null hypothesis that $\gamma_3^* = 0$).

A full analysis of the fit of the reduced model

$$X_t = \gamma_0 + \gamma_1^* \cos(\omega t) + \gamma_2^* \sin(\omega t) + \gamma_3^* \sin(2\omega t) + \epsilon_t. \quad (8.9)$$

can be performed using the commands:

```
lung2b.lmod <- lm(alllung~cosmonth+sinmonth+sin2month)
plot(month, alllung, xlab="Month", ylab="Monthly number of lung disease deaths", ylim=
c(1000,4000), xlim=c(1,12), xaxp=c(0,12,12), yaxp=c(1000,4000,3), cex.axis=1.2, cex.lab=1.4, pch=
16, cex=1.2, lines(predict(lung2b.lmod), lwd=2))
lung2b.pred <- fitted(lung2b.lmod) ; lung2b.sres <- studres(lung2b.lmod)
plot(lung2b.pred, lung2b.sres, xlab="Predicted value", ylab="Studentized residual", cex.axis=1.2,
cex.lab=1.4, pch=16, cex=1.2)
shapiro.test(lung2b.sres)
lung2b.sresb <- c(lung2b.sres[1:25],NA,lung2b.sres[26:60],NA,lung2b.sres[61:70])
bartlett.test(lung2b.sresb, month)
summary(lung2b.lmod)
```

The scatterplots produced for the data and fitted values, and the studentized residuals against the fitted values, appear in Fig. 8.5. Neither plot is indicative of any serious lack of fit. The p-values of the Shapiro–Wilk and Bartlett tests are 0.6419 and 0.0902, respectively, and hence there is no evidence against the normality and constant variance assumptions. From the summary of the model, all of its terms are very highly significant (largest p-value is 0.0004 in individual t-tests), and the parameter estimates are $\hat{\gamma}_0 = 2125.12$, $\hat{\gamma}_1^* = 454.18$, $\hat{\gamma}_2^* = 601.96$, $\hat{\gamma}_3^* = 108.69$ and $\hat{\sigma}^2 = 171.3$. The coefficient of determination (R^2) is 0.9093, and the adjusted R^2, 0.9052. Thus, the fitted model explains around 91%

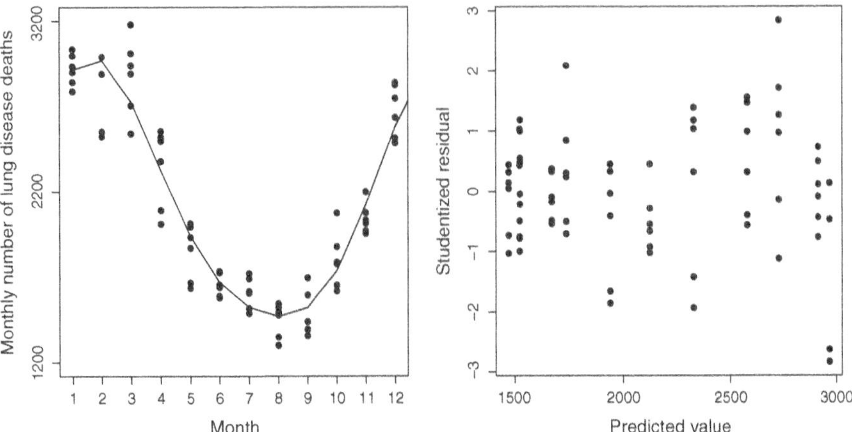

Figure 8.5 Scatterplots of: left, the monthly numbers of lung disease deaths with straight lines connecting the fitted values for model (8.9); right, the studentized residuals against the fitted values for the same fit

of the variation in the linear variable. From a consideration of the left-hand scatterplot in Fig. 8.5, it would appear that the shape of the fitted model around the trough is flatter than its shape around the peak. Whilst the various summaries quoted above suggest a good fit of (8.9) to the reduced data set, it must be remembered that two of the February observations were omitted when fitting the model.

8.4.3 *Skew Cosine Regression Model*

A model for linear–circular regression problems manifesting skewness is discussed by Batschelet (1981, Section 8.3), and is directly related to the unimodal asymmetric transformation of argument distributions introduced in Section 4.3.12. This model takes the form

$$X_t = \gamma_0 + \gamma_1 \cos((\omega t - \phi) + \nu \cos(\omega t - \phi)) + \epsilon_t, \tag{8.10}$$

ω again being the angular frequency, and γ_0, γ_1, ϕ and ν being parameters of the model requiring estimation. The overall level of X_t is controlled by γ_0, and the amplitude of the oscillations about that level by γ_1. For a given value of ν, ϕ determines where the peak occurs. For $\nu \in [-\pi/6, \pi/6]$, ν acts as a skewness parameter, its sign determining the direction of skewness. Clearly, when $\nu = 0$, (8.10) reduces to (8.5), which is symmetric. When ν takes values outside $[-\pi/6, \pi/6]$ its role as a skewness parameter is lost, as then it controls other shape characteristics of the model, including its modality.

When $\nu \neq 0$, nonlinear least-squares regression must be used to fit (8.10). Thus, in **R**, we make use of the **nls**, rather than the **lm**, function to fit it. For the monthly lung disease deaths, with the February values for 1976 and 1979 replaced by missing values, this can be done using, in addition to the commands used in the previous subsection, the extra commands:

```
lung.nlmod <- nls(alllung ~ gam0+gam1*cos(omega*month-phi+nu*cos(omega*month-phi)),
start=list(gam0=2000, gam1=1000, phi=1, nu=0))
summary(lung.nlmod)
```

Running the above code, the parameter estimates returned are $\hat{\gamma}_0 = 2122$, $\hat{\gamma}_1 = 750$, $\hat{\phi} = 0.93$ and $\hat{\nu} = 0.085$. However, the null hypothesis of $\nu = 0$ is not rejected by a t-test (p-value = 0.313), and hence, according to this analysis, the best fitting case of (8.10) is that with $\nu = 0$, i.e. model (8.5) fitted in Section 8.4.1. According to this analysis, then, there appears to be no appreciable skewness in the seasonal pattern of the lung disease deaths.

8.4.4 Symmetric Flat-Topped and Sharply Peaked Cosine Regression Model

In Section 8.4.2, the fitted model appeared to be flatter around the trough than around the peak. An alternative model capable of describing such departures from sinusoidal oscillations is also discussed by Batschelet (1981, Section 8.3). It is directly related to the unimodal symmetric transformation of argument distributions considered in Section 4.3.10, taking the form

$$X_t = \gamma_0 + \gamma_1 \cos((\omega t - \phi) + \lambda \sin(\omega t - \phi)) + \epsilon_t. \tag{8.11}$$

The roles of the parameters γ_0, γ_1, ω and ϕ are analogous to those for (8.10). Now, however, $\lambda \in [-\pi/3, \pi/3]$ is a parameter that regulates the shape of the symmetric model around the peak and trough. Positive λ values correspond to periodic functions that are more peaked than sinusoidal around the peak and flatter than sinusoidal around the trough. The opposite holds for negative values of λ. The basic cosine model (8.5) is recovered when $\lambda = 0$. Values of λ outside the interval $[-\pi/3, \pi/3]$ introduce secondary peaks and troughs.

In **R**, fitting (8.11) to the monthly lung disease deaths, with the February values for 1976 and 1979 replaced by missing values, is very similar to fitting (8.10). An extended analysis can be performed using the commands:

```
lung.nlmod2 <- nls(alllung~gam0+gam1*cos(omega*month-phi+lambda*sin(omega*month-phi)),
start=list(gam0=2000,gam1=1000,phi=1,lambda=0))
lungnlm2.pred <- fitted(lung.nlmod2) ; lungnlm2.res <- residuals(lung.nlmod2)
shapiro.test(lungnlm2.res)
lungnlm2.resb <- c(lungnlm2.res[1:25], NA, lungnlm2.res[26:60], NA, lungnlm2.res[61:70])
bartlett.test(lungnlm2.resb, month) ; summary(lung.nlmod2)
```

From the summary of the model, all of its terms are very highly significant (largest p-value is 0.0016), and hence the fitted model differs significantly from (8.5). The parameter estimates are $\hat{\gamma}_0 = 2224$, $\hat{\gamma}_1 = 762$, $\hat{\phi} = 0.92$, $\hat{\lambda} = 0.26$ and $\hat{\sigma}^2 = 173.5$. The positive value of $\hat{\lambda}$ squares with the findings for the fit of model (8.9), reflecting that around the peak the fitted model is more peaked than a sinusoid, and around the trough it is flatter. Analogous scatterplots to those on Figs. 8.4 and 8.5 display no obvious lack of fit, and the p-values for the Shapiro–Wilk and Bartlett tests are 0.9203 and 0.1193, respectively. As the scatterplot

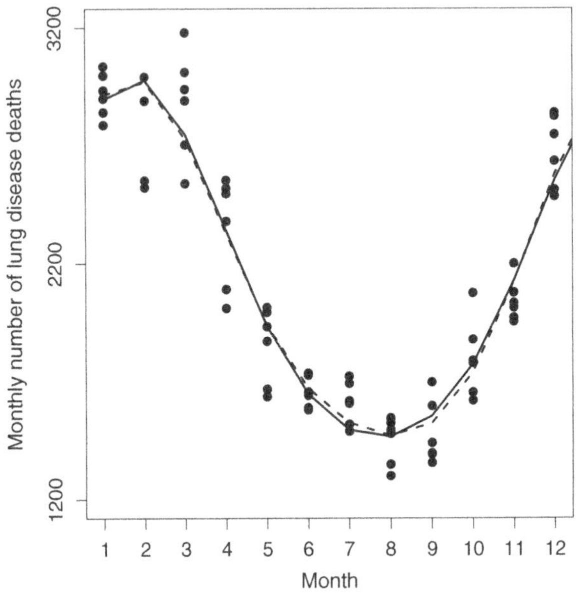

Figure 8.6 Scatterplot of the monthly numbers of lung disease deaths with solid straight lines connecting the fitted values for model (8.11) and dashed straight lines connecting the fitted values for model (8.9)

in Fig. 8.6 shows, the fitted values for this model and model (8.9) are very similar. Comparing all the various summaries for their respective fits, on purely statistical grounds there is very little to choose between them. From the point of view of interpretation, model (8.11) is certainly the more appealing of the two.

8.5 Regression for a Circular Response and Linear Regressors

In this section we consider the regression scenario in which we wish to model a circular response variable, Θ, as a function of the values taken by a number of linear explanatory variables, $X_1, \ldots, X_k$.

The **circular** package has a function **lm.circular** which allows us to explore the fit of a certain rather restrictive model. Specifically, suppose we have a set of n independent observations, $(\theta_1, \mathbf{x}_1), \ldots, (\theta_n, \mathbf{x}_n)$, where $\mathbf{x}_j$ denotes the vector $(x_{1j}, \ldots, x_{kj})$. The θ_j are assumed to have been drawn from a von Mises distribution with mean direction μ_j and concentration parameter κ, with μ_j being related to the explanatory variables $X_1, \ldots, X_k$ according to the equation

$$\mu_j = \mu + g(\gamma_1 X_1 + \cdots + \gamma_k X_k),$$

where μ and $\gamma_1, \ldots, \gamma_k$ are constants that (along with κ) must be estimated. For this model, the link function

$$g(u) = 2\tan^{-1}(u)$$

is generally used. For this choice of link function, $g(0) = 0$ (so μ can be interpreted as the origin), and $g(u)$ takes values between $-\pi$ and π as u ranges from $-\infty$ to ∞. The function **lm.circular** uses maximum likelihood estimation to estimate the unknown parameters. Details of the underlying theory can be found in Fisher and Lee (1992) and Fisher (1993, Section 6.4). This section draws very heavily on these two publications, which together provide the most coherent and thorough body of work relating to this problem.

We also use an example from Fisher (1993, Section 6.4), for which we attempt to model the direction of motion of periwinkles after being experimentally transplanted downshore as a function of the distance travelled by them after release. The data for the 31 periwinkles involved are presented in Fig. 8.7. The sea was located at a bearing of 275 degrees from the release point, and the data seem to suggest that those travelling in a direction towards the sea travelled shorter distances than those travelling in directions away from it. From a biological perspective, this could have been due to a propensity of the periwinkles to stop moving once they are close to the sea. We can fit the regression model to the periwinkle data using the code:

```
distance <- c(107,46,33,67,122,69,43,30,12,25,37,69,5,83,68,38,21,1,71,60,71,71,57,53,38,70,7,48,
7,21,27)
directdeg <- c(67,66,74,61,58,60,100,89,171,166,98,60,197,98,86,123,165,133,101,105,71,84,75,98,
83,71,74,91,38,200,56)
cdirect <- circular(directdeg*2*pi/360)
lm.circular(type="c-l", y=cdirect, x=distance, init=0.0)
```

Notice that we must specify the type of regression, here *circular–linear*, since the function **lm.circular** can also be used for *circular–circular* regression (see Section 8.6). We must also specify a starting estimate for γ_1. This code produces estimated values and standard errors for μ, κ and γ_1, and the p-value for an hypothesis test of H_0: $\gamma_1 = 0$. The estimates returned are $\hat{\mu} = 2.43$ radians, $\hat{\kappa} = 3.22$ and $\hat{\gamma}_1 = -0.008$ (s.e. = 0.001, p-value < 0.001). However, several notes of caution must be emphasized at this point. Firstly, for our example data, the function's iterative procedure fails to converge for initial values that deviate far from zero. This can be understood in this particular case where a very sharp local maximum of the log-likelihood exists just to the left of zero, but otherwise the dominant features of the log-likelihood function are two other maxima at $\pm\infty$ (see Fig. 6.11 of Fisher (1993)). Fisher (1993, Section 6.4) warns that such complex log-likelihood surfaces are commonplace for this sort of model fitting.

In many cases, including our example, the linear predictor may well influence the concentration as well as the mean direction of the distribution of the circular response variable. Fisher (1993, Section 6.4) discusses how to fit such a mixed model. However the challenges of model fitting are even greater for this more complex case. Such model fitting deserves a book in its own right, since the most appropriate method of attack will depend both on the details of each problem and the previous experience and preferences of each investigator.

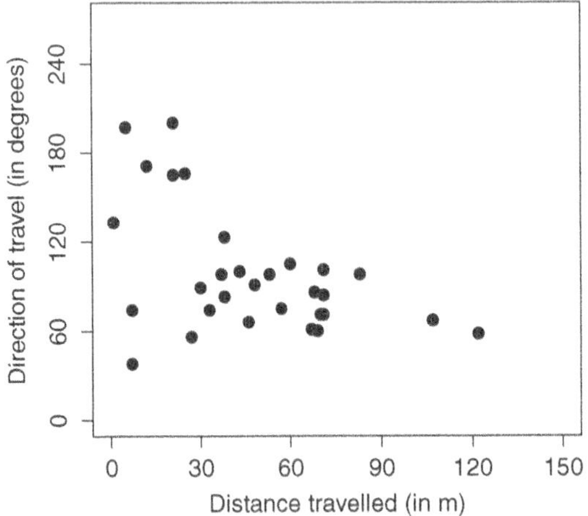

Figure 8.7 Scatterplot of directions and distances travelled by 31 periwinkles after being experimentally translocated downshore

All we can add is that Fisher (1993, Section 6.4) remains the clearest and most thorough investigation of this theoretically, but armed with that theory the challenge becomes one of conventional (if far from trivial) mixed model fitting.

8.6 Regression for a Circular Response and a Circular Regressor

The situation in which a circular random variable, Θ, is modelled as a function of the value taken by another, Ψ, has received very little attention. Suppose observations $(\theta_1, \psi_1), \ldots, (\theta_n, \psi_n)$ on (Θ, Ψ) are available. Jammalamadaka and SenGupta (2001) recommend fitting the general linear model

$$\cos(\Theta_j) = \gamma_0^c + \sum_{k=1}^{m} \left(\gamma_{ck}^c \cos(k\psi_j) + \gamma_{sk}^c \sin(k\psi_j)\right) + \epsilon_{1j},$$
$$\sin(\Theta_j) = \gamma_0^s + \sum_{k=1}^{m} \left(\gamma_{ck}^s \cos(k\psi_j) + \gamma_{sk}^s \sin(k\psi_j)\right) + \epsilon_{2j}, \quad (8.12)$$

where, for the γ's (the regression coefficients), the c and s superindices relate to $\cos(\Theta_j)$ and $\sin(\Theta_j)$, respectively, the ck and sk subindices relate to $\cos(k\psi_j)$ and $\sin(k\psi_j)$, respectively, and $(\epsilon_{1j}, \epsilon_{2j})$ is an error vector with mean $(0, 0)$ and whose dispersion matrix must be estimated from the data. This model can also be fitted using **R**'s **lm.circular** function.

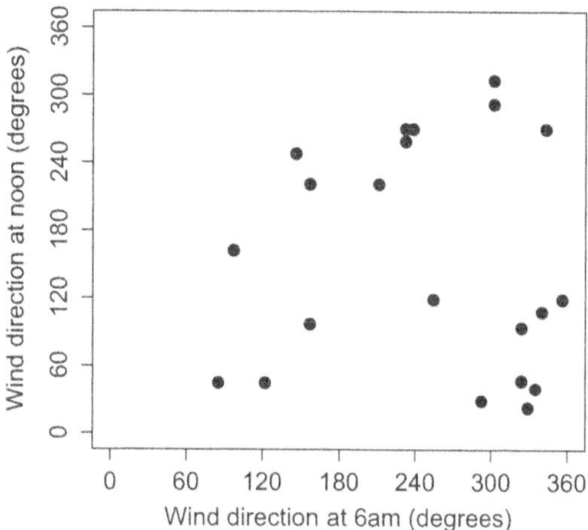

Figure 8.8 Wind directions measured at a weather station in Milwaukee at 6am and noon on 21 consecutive days

To illustrate the use of the **lm.circular** function in this context, we apply it in the analysis of a data set first published by Johnson and Wehrly (1977) on the direction of the wind at a weather station in Milwaukee measured at 6am (which we will denote by Ψ) and at noon (denoted by Θ) on each of 21 consecutive days. The data are reproduced by Jammalamadaka and SenGupta (2001, page 179), and are represented graphically in Fig. 8.8. There appears to be some suggestion of an association. If we adopt the approach recommended in Section 8.3.1 for testing for rotational dependence, we obtain $r_{FL} = 0.19$, with, when we ran the randomization test, an associated p-value of 0.018. We can fit model (8.12) with $m = 1$ to these data using the commands:

```
psideg <- c(356,97,211,232,343,292,157,302,335,302,324,85,324,340,157,238, 254,146,232,122,329)
thetadeg <- c(119,162,221,259,270,29,97,292,40,313,94,45,47,108,221,270,119, 248,270,45,23)
cpsirad <- circular(psideg*2*pi/360) ; cthetarad <- circular(thetadeg*2*pi/360)
lm.circular(type="c-c", y=cthetarad, x=cpsirad, order=1)
```

Notice that we must convert the data into variables of the type *circular* in radians. We must also specify that we are performing circular–circular regression with **type = "c-c"**, and we must specify the order, m, to use when fitting model (8.12). A justification for the use of $m = 1$ for this example is given below. **R** returns various results including the fitted values, residuals and estimated regression coefficients. It also generates p-values for testing whether the two extra regression coefficients (the γ's) associated with $\cos(\Theta_i)$ for a model of order $m + 1$ (rather than m), and their analogues associated with $\sin(\Theta_i)$, are significantly different from zero. In this case the p-value for the two extra regression coefficients for $\cos(\Theta_i)$ is 0.0997, and that for the two extra for $\sin(\Theta_i)$ is 0.9475. Thus, for these data, there appears to be no reason to use a model of order 2 rather than one of order 1.

Finally, the function provides a measure (**rho**) which is the square root of the average of the squares of the estimated conditional concentration parameters of Θ given ψ. This can be interpreted as a measure of model fit equivalent to the square of Pearson's product moment coefficient in relation to ordinary least squares. It too takes values between 0 and 1, with increasing values suggesting a better fit. For this example, it takes the value 0.502, which suggests that the fit of the model is moderately good.

8.7 Multivariate Regression with Circular Regressors

Some recent papers have begun to explore multivariate regression where one or more of the regressor variables is circular. This is not an area that we could tackle with brevity, since the conceptual and computational challenges are considerable. For interested readers, recent relevant literature sources include Lund (1999, 2002), Bhattacharya and SenGupta (2009), and Qin *et al.* (2011).

Appendix

FURTHER READING

1 Books on Circular Statistics

Only six books have previously been published which treat circular statistics in depth. In order of publication, they are: Mardia (1972), Batschelet (1981), Upton and Fingleton (1989), Fisher (1993), Mardia and Jupp (1999) and Jammalamadaka and SenGupta (2001). The first three are out of print but university libraries may well have copies of them. All six books have extensive bibliographies which, between them, provide an excellent introduction to the published literature in the field.

Mardia and Jupp (1999) is a heavily revised, updated and extended version of Mardia (1972), which, together with Jammalamadaka and SenGupta (2001), provide the most theoretical treatments of the subject. For theoreticians, Mardia (1972) is still the best introduction to the subject. Batschelet (1981), Upton and Fingleton (1989) and Fisher (1993) generally use a more applications-based style of presentation, and after introducing concepts and theory provide details of how to apply a given method together with an example of its use.

All but Fisher (1993) show their age in the sense that they generally assume that users will perform their calculations using a hand calculator. Jammalamadaka and SenGupta (2001) is the only one of them that includes any code, but the S-Plus based CircStats package to which it refers is of rather limited scope. We have used a style more akin to that employed in Batschelet (1981), Upton and Fingleton (1989) and Fisher (1993) but have included R code with which to apply techniques.

In terms of its statistical content, the present book has much in common with: Chapters 1–3 and 7 of Mardia (1972); Chapters 1–6, 8–9 and 15–17 of Batschelet (1981); Chapter 9 of Upton and Fingleton (1989); Chapters 1–6 and 8 of Fisher (1993); Chapters 1–3, 6 and 8 of Mardia and Jupp (1999); and Chapters 1–2 and 5–8 of Jammalamadaka and SenGupta (2001). The methodology we have considered is entirely frequentist in nature. We have not considered, for example, robust methods nor Bayesian methods. Aspects of these two approaches are considered in Mardia and Jupp (1999) and Jammalamadaka and SenGupta (2001) but neither is particularly well-developed in the context of circular statistics. Neither have we considered outlier detection (which is far less of a problem for circular data than for linear data because the unit circle is a compact support), time series analysis or spatial analysis for circular data. The last two forms of analysis are covered by Fisher (1993, Chapter 7).

2 Internet-based Resources

Searching for the topic 'circular statistics' using search engines like *Scirus* and *Google Scholar*, or publication databases such as the *Web of Knowledge* and *Scopus*, is an efficient way of identifying webpages which make reference to circular statistics and the latest journal articles dealing with applications and theoretical developments in the field.

REFERENCES

Numbers in square brackets refer to pages where each reference is used.

Abe, T. and Pewsey, A. (2011*a*). Sine-skewed circular distributions. *Statistical Papers*, **52**, 683–707. [65]

Abe, T. and Pewsey, A. (2011*b*). Symmetric circular models through duplication and cosine perturbation. *Computational Statistics and Data Analysis*, **55**, 3271–82. [76]

Abe, T., Pewsey, A., and Shimizu, K. (2013). Extending circular distributions through transformation of argument. *Annals of the Institute of Statistical Mathematics*. DOI 10.1007/s10463-012-0394-5. [64, 67, 123]

Abe, T., Shimizu, K., and Pewsey, A. (2009). On Papakonstantinou's extension of the cardioid distribution. *Statistics and Probability Letters*, **79**, 2138–47. [64]

Adler, J. (2010). *R in a Nutshell* (3rd edn). O'Reilly Media, Sebastopol. [4]

Agostinelli, C. (2007). Robust estimation for circular data. *Computational Statistics and Data Analysis*, **51**, 5867–75. [11]

Akaike, H. (1974). A new look at the statistical model identification. *IEEE Transactions on Automatic Control*, **19**, 716–23. [114]

Aneshansley, D. J. and Larkin, T. S. (1981). *V*-test is not a statistical test of 'homeward' direction. *Nature*, **293**, 239. [86]

Azzalini, A. (1985). A class of distributions which includes the normal ones. *Scandinavian Journal of Statistics*, **12**, 171–8. [65]

Bartels, R. (1984). Estimation in a bidirectional mixture of von Mises distributions. *Biometrics*, **40**, 777–84. [76]

Batschelet, E. (1981). *Circular Statistics in Biology*. Academic Press, London. [27, 64, 67, 82, 145, 164, 165, 171]

Beran, R. (1979). Exponential models for directional data. *Annals of Statistics*, **7**, 1162–78. [75, 76]

Berens, P. (2009). CircStat: A Matlab toolbox for circular statistics. *Journal of Statistical Software*, **31**, 1–21. [6]

Best, D. and Fisher, N. (1979). Efficient simulation of the von Mises distribution. *Applied Statistics*, **28**, 152–7. [57]

Bhattacharya, S. and SenGupta, A. (2009). Bayesian analysis of semiparameteric linear–circular models. *Journal of Agricultural, Biological and Environmental Statistics*, **14**, 33–65. [170]

Bogdan, M., Bogdan, K., and Futschik, A. (2002). A data driven smooth test for circular uniformity. *Annals of the Institute of Statistical Mathematics*, **54**, 29–44. [82]

Boomsma, W., Mardia, K. V., Taylor, C. C., Ferkinghoff-Borg, J., Krogh, A., and Hamelryck, T. (2008). A generative, probabilistic model of local protein structure. *Proceedings of the National Academy of Sciences of the United States of America*, **105**, 8932–7. [2]

Brown, B. M. (1994). Grouping corrections for circular goodness-of-fit tests. *Journal of the Royal Statistical Society B*, **56**, 275–83. [83]

Byrd, R. H., Lu, P., Nocedal, J., and Zhu, C. (1995). A limited memory algorithm for bound constrained optimization. *SIAM Journal on Scientific Computing*, **16**, 1190–208. [108]

Cartwright, D. E. (1963). The use of directional spectra in studying the output of a wave recorder on a moving ship. In *Ocean Wave Spectra* (ed.), pp. 203–18. Prentice Hall, Englewood Cliffs, NJ. [50, 52]

Choulakian, V., Lockhart, R. A., and Stephens, M. A. (1994). Cramér–von Mises statistics for discrete distributions. *Canadian Journal of Statistics*, **22**, 125–37. [83]

Cox, D. R. (1975). Contribution to discussion of Mardia (1975), Statistics of directional data. *Journal of the Royal Statistical Society B*, **37**, 380–1. [76]

Cox, D. R. and Lewis, P. A. W. (1966). *The Statistical Analysis of Series of Events*. Methuen, London. [80]

Crawley, M. J. (2012). *The R Book* (2nd edn). John Wiley, Chichester. [4]

Dryden, I. L. and Mardia, K. V. (1998). *Statistical Shape Analysis*. John Wiley, Chichester. [7]

Edwards, J. H. (1961). The recognition and estimation of cyclic trends. *Annals of Human Genetics*, **25**, 83–7. [82]

Efron, B. (1979). Bootstrap methods: another look at the jackknife. *Annals of Statistics*, **7**, 1–26. [88]

Ekstrom, C. T. (2011). *The R Primer*. CRC Press, Boca Raton. [4]

Fernández-Durán, J. J. (2004). Circular distributions based on nonnegative trigonometric sums. *Biometrics*, **60**, 499–503. [76]

Fisher, N. I. (1986). Robust comparison of dispersion for samples of directional data. *Australian Journal of Statistics*, **28**, 213–9. [139]

Fisher, N. I. (1993). *Statistical Analysis of Circular Data*. Cambridge University Press, Cambridge. [2, 23, 32, 50, 79, 80, 98, 99, 102, 131, 132, 134, 135, 136, 137, 139, 141, 153, 156, 167, 168, 171]

Fisher, N. I. and Hall, P. G. (1991). Bootstrap algorithms for small samples. *Journal of Statistical Planning and Inference*, **27**, 157–69. [135]

Fisher, N. I. and Lee, A. J. (1981). Nonparametric measures of angular–linear association. *Biometrika*, **68**, 629–36. [153]

Fisher, N. I. and Lee, A. J. (1983). A correlation coefficient for circular data. *Biometrika*, **70**, 327–32. [154, 157]

Fisher, N. I. and Lee, A. J. (1992). Regression models for an angular response. *Biometrics*, **48**, 665–77. [167]

Fisher, N. I., Lewis, T., and Embleton, B. J. J. (1993). *Statistical Analysis of Spherical Data*. Cambridge University Press, Cambridge. [6]

Freedman, L. S. (1979). The use of a Kolmogorov–Smirnov type statistic in testing hypotheses about seasonal variation. *Journal of Epidemiology and Community Health*, **33**, 223–8. [83]

Freedman, L. S. (1981). Watson's U_n^2 statistic for a discrete distribution. *Biometrika*, **68**, 708–11. [83]

Gatto, R. and Jammalamadaka, S. R. (2007). The generalized von Mises distribution. *Statistical Methodology*, **4**, 341–53. [76]

Gradshteyn, I. S. and Ryzhik, I. M. (1994). *Table of Integrals, Series, and Products* (5th edn). Academic Press, San Diego. [58]

Hall, P., Watson, G. S., and Cabrera, J. (1987). Kernel density estimation with spherical data. *Biometrika*, **74**, 751–62. [16]

Hand, D. J., Daly, F., Lunn, A. D., McConway, K. J., and Ostrowski, E. (1994). *Handbook of Small Data Sets*. Chapman and Hall, New York. [160]

Jammalamadaka, S. R. and Kozubowski, T. J. (2004). New families of wrapped distributions for modeling skew circular data. *Communications in Statistics—Theory and Methods*, **33**, 2059–74. [75]

Jammalamadaka, S. R. and Sarma, Y. R. (1988). A correlation coefficient for angular variables. In *Statistical Theory and Data Analysis II* (ed. K. Matusita), pp. 349–64. North Holland, Amsterdam. [157]

Jammalamadaka, S. R. and SenGupta, A. (2001). *Topics in Circular Statistics*. World Scientific, Singapore. [6, 32, 46, 75, 157, 168, 169, 171]

Jeffreys, H. (1948). *Theory of Probability* (2nd edn). Oxford University Press, Oxford. [48]

Johnson, R. A. and Wehrly, T. E. (1977). Measures and models for angular correlation and angular–linear correlation. *Journal of the Royal Statistical Society B*, **39**, 222–9. [77, 150, 169]

Johnson, R. A. and Wehrly, T. E. (1978). Some angular–linear distributions and related regression models. *Journal of the American Statistical Association*, **73**, 602–6. [78]

Jones, M. C., Kato, S., and Pewsey, A. (2013). On a class of circulas. To appear. [77]

Jones, M. C. and Pewsey, A. (2005). A family of symmetric distributions on the circle. *Journal of the American Statistical Association*, **100**, 1422–8. [62, 108, 113, 123]

Jones, M. C. and Pewsey, A. (2012). Inverse Batschelet distributions for circular data. *Biometrics*, **68**, 183–93. [70, 72, 123]

Jupp, P. E. and Mardia, K. V. (1989). A unified view of the theory of directional statistics. *International Statistical Review*, **57**, 261–94. [153]

Jupp, P. E. and Spurr, B. D. (1983). Sobolev tests for symmetry of directional data. *Annals of Statistics*, **11**, 1225–31. [86]

Jupp, P. E. and Spurr, B. D. (1985). Sobolev tests for independence of directions. *Annals of Statistics*, **13**, 1140–55. [158]

Kabacoff, R. I. (2011). *R in Action: Data Analysis and Graphics with R*. Manning, Shelter Island, NY. [4]

Kato, S. and Jones, M. C. (2010). A family of distributions on the circle with links to, and applications arising from, Möbius transformation. *Journal of the American Statistical Association*, **105**, 249–62. [46, 75]

Kato, S. and Pewsey, A. (2013). A Möbius transformation-induced distribution on the torus. To appear. [77]

Kuiper, N. H. (1960). Tests concerning random points on the circle. *Koninklijke Nederlandse Akademie Van Wetenschappen, Proceedings Series A*, **63**, 38–47. [145]

Lockhart, R. A. and Stephens, M. A. (1985). Tests of fit for the von Mises distribution. *Biometrika*, **72**, 647–52. [105]

Lund, U. (1999). Least circular distance regression for directional data. *Journal of Applied Statistics*, **26**, 723–33. [170]

Lund, U. (2002). Tree-based regression for a circular response. *Communications in Statistics—Theory and Methods*, **31**, 1549–60. [170]

Maag, U. R. (1966). A k-sample analogue of Watson's U^2 statistic. *Biometrika*, **53**, 579–83. [145]

Maksimov, V. M. (1967). Necessary and sufficient statistics for the family of shifts of probability distributions on continuous bicompact groups (in Russian). *Theoria Verojatna*, **12**, 307–21. [76]

Manly, B. F. J. (2007). *Randomisation, Bootstrap and Monte Carlo Methods in Biology* (3rd edn). Chapman and Hall, Boca Raton. [9]

Mardia, K. V. (1972). *Statistics of Directional Data*. Academic Press, London. [30, 31, 46, 75, 118, 142, 171]

Mardia, K. V. (1975). Statistics of directional data (with discussion). *Journal of the Royal Statistical Society B*, **37**, 349–93. [77]

Mardia, K. V. (1976). Linear–circular correlation coefficients and rhythometry. *Biometrika*, **63**, 403–5. [150, 152]

Mardia, K. V. and Jupp, P. E. (1999). *Directional Statistics*. John Wiley, Chichester. [2, 6, 14, 44, 46, 47, 55, 58, 82, 83, 90, 150, 152, 153, 158, 171]

Mardia, K. V. and Sutton, T. W. (1975). On the modes of a mixture of two von Mises distributions. *Biometrika*, **62**, 699–701. [76]

Mardia, K. V. and Sutton, T. W. (1978). A model for cylindrical variables with applications. *Journal of the Royal Statistical Society B*, **40**, 229–33. [78]

Mardia, K. V., Taylor, C. C., and Subramaniam, G. K. (2007). Protein bioinformatics and mixtures of bivariate von Mises distributions for angular data. *Biometrics*, **63**, 505–12. [2]

McCullagh, P. (1996). Möbius transformation and Cauchy parameter estimation. *Annals of Statistics*, **24**, 787–808. [54]

Modarres, R. and Gastwirth, J. L. (1996). A modified runs test for symmetry. *Statistics and Probability Letters*, **31**, 107–12. [87]

Moore, B. R. (1980). A modification of the Rayleigh test for vector data. *Biometrika*, **67**, 175–80. [146]

Morellato, L. P. C., Alberti, L. F., and Hudson, I. L. (2010). Application of circular statistics in plant phenology: a case studies approach. In *Phenological Research: Methods for Environmental and Climate Change Analysis* (ed. I. L. Hudson and M. R. Keatley), pp. 339–59. Springer, Dordrecht. [2, 6]

Neuhäuser, M. (2012). *Nonparametric Statistical Tests: A Computational Approach*. CRC, Boca Raton. [9]

Oliveira, M., Crujeiras, R. M., and Rodríguez-Casal, A. (2012). A plug-in rule for bandwidth selection in circular density estimation. *Computational Statistics and Data Analysis*, **56**, 3898–908. [17]

Pabst, B. and Vicentini, H. (1978). Dislocation experiments in the migrating seastar *Astropecten jonstoni*. *Marine Biology*, **48**, 271–8. [23]

Papakonstantinou, V. (1979). *Beiträge zur zirkulären Statistik*. Ph.D. thesis, University of Zurich, Switzerland. [64, 67]

Pewsey, A. (2000). The wrapped skew-normal distribution on the circle. *Communications in Statistics—Theory and Methods*, **29**, 2459–72. [75]

Pewsey, A. (2002*a*). *Contributions to the Analysis of Skew Data on the Line and Circle*. Ph.D. thesis, Open University, Milton Keynes, UK. [75]

Pewsey, A. (2002*b*). Testing circular symmetry. *Canadian Journal of Statistics*, **30**, 591–600. [87, 102]

Pewsey, A. (2004*a*). The large-sample joint distribution of key circular statistics. *Metrika*, **60**, 25–32. [7, 42, 90]

Pewsey, A. (2004*b*). Testing for circular reflective symmetry about a known median axis. *Journal of Applied Statistics*, **31**, 575–85. [87]

Pewsey, A. (2006). Modelling asymmetrically distributed circular data using the wrapped skew-normal distribution. *Environmental and Ecological Statistics*, **13**, 257–69. [75]

Pewsey, A. (2008). The wrapped stable family of distributions as a flexible model for circular data. *Computational Statistics and Data Analysis*, **52**, 1516–23. [75]

Pewsey, A., Lewis, T., and Jones, M. C. (2007). The wrapped *t* family of circular distributions. *Australian and New Zealand Journal of Statistics*, **49**, 79–91. [56, 75]

Pewsey, A., Shimizu, K., and de la Cruz, R. (2011). On an extension of the von Mises distribution due to Batschelet. *Journal of Applied Statistics*, **38**, 1073–85. [64]

Pycke, J.-R. (2010). Some tests for uniformity of circular distributions powerful against multi-modal alternatives. *Canadian Journal of Statistics*, **38**, 80–96. [82]

Qin, X., Zhang, J.-S., and Yan, X.-D. (2011). A nonparametric circular–linear multivariate regression model with a rule-of-thumb bandwidth selector. *Computers and Mathematics with Applications*, **62**, 3048–55. [170]

Reed, W. J. and Pewsey, A. (2009). Two nested families of skew-symmetric circular distributions. *Test*, **18**, 516–28. [75]

Rothman, E. D. (1971). Tests of coordinate independence for a bivariate sample on the torus. *Annals of Mathematical Statistics*, **42**, 1962–9. [158]

Ruhkin, A. L. (1972). Some statistical decisions about distributions on a circle for large samples. *Sankhya A*, **34**, 243–50. [76]

Schach, S. (1969). Nonparametric symmetry tests for circular distributions. *Biometrika*, **56**, 571–7. [86]

Schmidt-Koenig, K. (1963). On the role of the loft, the distance and the site of release in pigeon homing (the 'cross-loft experiment'). *Biological Bulletin*, **125**, 154–64. [99]

Schwarz, G. E. (1978). Estimating the dimension of a model. *Annals of Statistics*, **6**, 461–4. [114]

Self, S. G. and Liang, K.-Y. (1987). Asymptotic properties of maximum likelihood estimators and likelihood ratio tests under nonstandard conditions. *Journal of the American Statistical Association*, **82**, 605–10. [113]

Spurr, B. D. (1981). On estimating the parameters in mixtures of circular normal distributions. *Mathematical Geology*, **13**, 163–73. [76]

Spurr, B. D. and Koutbeiy, M. A. (1991). A comparison of various methods for estimating the parameters in mixtures of von Mises distributions. *Communications in Statistics—Simulation and Computation*, **20**, 725–41. [76]

Steele, M. and Chaseling, J. (2006). Powers of discrete goodness-of-fit test statistics for a uniform null against a selection of alternative distributions. *Communications in Statistics—Simulation and Computation*, **35**, 1067–75. [83]

Stephens, M. A. (1969). A goodness-of-fit statistic for the circle, with some comparisons. *Biometrika*, **56**, 161–8. [82]

Stephens, M. A. (1972). Multisample tests for the von Mises distribution. *Journal of the American Statistical Society*, **67**, 456–61. [136]

Taylor, C. C. (2008). Automatic bandwidth selection for circular density estimation. *Computational Statistics and Data Analysis*, **52**, 3493–500. [16]

Umbach, D. and Jammalamadaka, S. R. (2009). Building asymmetry into circular distributions. *Statistics and Probability Letters*, **79**, 659–63. [65]

Upton, G. J. G. and Fingleton, B. (1989). *Spatial Data Analysis by Example*, Volume 2: Categorical and Directional Data. John Wiley, New York. [23, 82, 145, 154, 156, 171] von Mises, R. (1918). Über die 'Ganzzahligkeit' der Atomgewichte und verwandte Fragen. *Physikalische Zeitschrift*, **19**, 490–500. [58]

Wallraff, H.G. (1979). Goal-orientated and compass-oriented movements of displayed homing pigeons after confinement in differentially shielded aviaries. *Behavioral Ecology and Scoiobiology*, **5**, 201–25. [139]

Watson, G. S. (1962). Goodness-of-fit tests on a circle II. *Biometrika*, **49**, 57–63. [144]

Watson, G. S. (1983). *Statistics on Spheres*. John Wiley, New York. [6, 134]

Watson, G. S. and Williams, E. J. (1956). On the construction of significance tests on the circle and on the sphere. *Biometrika*, **43**, 344–52. [81, 136]

Wehner, R. and Müller, M. (1985). Does interocular transfer occur in visual navigation by ants? *Nature*, **315**, 228–9. [131]

Wheeler, S. and Watson, G. S. (1964). A distribution-free two-sample test on the circle. *Biometrika*, **51**, 256–7. [142]

Yfantis, E. A. and Borgman, L. E. (1982). An extension of the von Mises distribution. *Communications in Statistics—Theory and Methods*, **11**, 1695–1706. [76]

Zar, J. H. (2010). *Biostatistical Analysis* (5th edn). Prentice Hall, New York. [145, 146, 147]

INDEX

D

The manufacturer's authorised representative in the EU for product safety is Oxford University Press España S.A. of El Parque Empresarial San Fernando de Henares, Avenida de Castilla, 2 - 28830 Madrid (www.oup.es/en or product.safety@oup.com). OUP España S.A. also acts as importer into Spain of products made by the manufacturer.

Printed and bound by CPI Group (UK) Ltd, Croydon, CR0 4YY

06/07/2026

02157632-0005